PVP-Vol. 492

PROCEEDINGS OF THE ASME

PRESSURE VESSELS AND PIPING DIVISION

– 2004 –

2004 FORUM ON FLUID TRANSIENTS

RECENT ADVANCES IN SOLIDS AND STRUCTURES

presented at
2004 ASME INTERNATIONAL MECHANICAL ENGINEERING CONGRESS AND EXPOSITION
NOVEMBER 13–19, 2004
ANAHEIM, CALIFORNIA USA

sponsored by
THE PRESSURE VESSELS AND PIPING DIVISION, ASME

A S M E
Three Park Avenue ◆ New York, N.Y. 10016

ISBN No. 0-7918-4718-7

CONTENTS

FORUM ON FLUID TRANSIENTS
INTRODUCTION

This forum presents five papers and two tutorials in three sessions on the broad subject of Fluid Transients. These sessions are sponsored by the Fluid Structure Interaction (FSI) Committee of the ASME Pressure Vessels and Piping (PVP) Division, and the Fluid Application and Systems Technical Committee (FASTC) of the Fluids Engineering Division (FED). They are part of the 2004 ASME International Mechanical Engineering Congress and Exposition (IMECE). The Forum on Fluid Transient has been a recurring annual theme at these IMECE meetings for several years.

Two of the five papers in the first session give the results of analytic investigations of the fluid structure interaction between thin cylindrical shells and fluid flowing through a narrow passage. This is for the cases of simply supported and freely supported shells. In the final installment of a six-part series of papers, the third paper compares the theoretical predictions of water hammer induced stresses in piping using the simplified dynamic stress theory to experimental measurements. The reduction of damaging vibrations in the recycle piping system on the Main Oil Line export pumps are discussed in the fourth paper. This reduction during the pump start/shut down transient is achieved through the modification of control valve trim. The fifth paper reports on the design optimization using a CFD model of the secondary airflows between individual burners in a fossil power plant. The CFD results are validated against the experimental data from a 1/8th scale model.

The first of the two tutorials presents simple models for quick predictions of basic fluid transients. Examples are included in this presentation involving fluid system response times, vibration characteristics, pressure and force magnitudes from fluid transients, and some industry surprises and failures that could have been avoided by a simple models of transient fluid phenomena. The second tutorial presentation includes live computer simulations with graphical display of various typical fluid transient problems. (Editorial Note: Tutorials are not included in the proceedings.)

The sessions chairs would like to take this opportunity to thank all the authors who have contributed to this forum. Thanks are also due to Mr. Artin Dermenjian and Dr. Howard Chung who helped in the organization of this Forum.

Asif H. Arastu
Bechtel Corporation
50 Beale Street, San Francisco, California 94105

Fred J. Moody
827 Larkspur Lane, Murphys, California 95247

Proceedings of IMECE04
2004 ASME International Mechanical Engineering Congress and Exposition
November 13-20, 2004, Anaheim, California USA

IMECE2004-59491

INSTABILITY OF AN AXIAL LEAKAGE FLOW-INDUCED VIBRATION OF THIN CYLINDRICAL SHELLS HAVING FREELY SUPPORTED END

Katsuhisa Fujita and Makoto Kato
Mechanical Systems Engineering, Graduate School of Engineering,
Osaka Prefecture University, 1-1, Gakuen-cho, Sakai-City, Osaka Prefecture, 599-8531, Japan
fujita@mecha.osakafu-u.ac.jp

ABSTRACT

When thin cylindrical shells having freely supported end at the downstream side such as heat-shielding shells of afterburners, labyrinth air seals, annular structures in large diameter pipings and valves are subjected to axial leakage flows, an unstable vibration and a fatigue failure are apt to be occurred. In this paper, the unstable vibration of thin cylindrical shells is analytically investigated considering the fluid structure interaction between shells and fluids flowing through a narrow passage. The coupled equation of motion between shells and fluids is derived using the Flügge's shell theory and the Navier-Stokes equation. Especially, focusing on the higher circumferential vibrations, the unstable phenomenon of thin cylindrical shells is clarified by using root locus based on the complex eigenvalue analysis by using the mode functions obtained by the exact solution based on the Flügge's shell theory. The influence of shell-dimensions and so forth on the threshold of the instability of the coupled vibration of shells and flowing fluids are investigated and discussed.

NOMENCLATURE

a: Radius of the inner thin cylindrical shell.
C: Extensional rigidity.
$C_{a,n}$: Added damping matrix.
D: Flexural rigidity.
H, $\bar{H}$, ΔH: Annulus gap, steady gap, unsteady gap.
h: Thickness of a shell.
$K_{s,n}$, $K_{a,n}$: Structural modal stiffness and added stiffness matrices.
ℓ: Length of a shell.
$M_{s,n}$, $M_{a,n}$: Structural modal mass and added mass matrices.
m: Axial mode number.
m_x, m_θ: Moments acting on a shell.
n: Circumferential mode number.
P, $\bar{P}$, ΔP: Pressure, steady pressure, unsteady pressure.
p_x, p_θ, p_z: Axial, circumferential, radial forces acting on a shell.
Q_x, $\bar{Q}_x$, ΔQ_x: Axial flow rate, steady axial flow rate, unsteady axial flow rate.
Q_y, ΔQ_y: Circumferential flow rate, unsteady circumferential flow rate.
t: Time.
U, V, W: Flow velocity in x-, θ-, z_f-directions ($y = a\theta$).
u, v, w: Displacement of a shell in x-, θ-, z_s-directions ($y = a\theta$).
$\hat{u}_{mn}$, $\hat{v}_{mn}$, $\hat{w}_{mn}$: Eigenmodes of a shell in axial, circumferential and radial directions.
x: Coordinate corresponding to axial direction of a shell.
y: Displacement corresponding to circumferential direction of a shell (y-axis is transformed by $\theta = y/a$).
z_f, z_s: Coordinate corresponding to radial direction of a fluid and a shell.
ζ_{in}, ζ_{out}: Pressure loss coefficients at inlet and outlet.
θ: Angle corresponding to circumferential direction (θ-axis is transformed by $y = a\theta$).
μ: $= \rho_s h$ (mass per unit area).
ν_s: Poisson's ratio.
ν_f: Kinematic viscosity.
$\{\xi_n(t)\}$: Displacement vector of a shell.
$\xi_{mn}(t)$: Modal displacement of a shell.
ρ_s, ρ_f: Density of a shell and a fluid.
ω_{mn}: Eigenfrequency in a vacuum.

INTORODUCTION

A fluid is often conducted as the energy transfer medium in many industrial machinery. On the other hand, many machines and structures must be constructed and operated in flowing fluid. The velocity of fluid becomes faster in order to increase the efficiency and make the size more compact. In this paper, the fluid-elastic instability which occurs in a coaxial cylindrical shell structures subjected to an axial leakage flow is investigated. The machines and structures which are composed of coaxial cylindrical shells are recognized in heat-shielding shells of afterburners, core internals of nuclear power plants, annular structures in large diameter pipings, labyrinth air seals and valves[(1)-(3)].

The vibrational mechanisms of machines and structures subjected to an axial leakage flow have been studied by many researchers[(4)-(11)]. However, the studies conducted to be based on a shell theory have not reported so many. Accordingly, the further remaining study can be considered to be still necessary for developing the improved coupled vibrational analysis methodology. In this paper, considering the fluid structure interaction between shells and fluids flowing through a narrow passage, the coupled equation of motion between shells and flowing fluids is derived using the Flügge's shell theory and the Navier-Stokes equation which is applied to an annular leakage flow in accompanying with some approximated assumptions. The vibrational characteristics of a thin cylindrical shell can be classified into the axisymmetric vibration, the lateral vibration and the circumferential vibration. Especially, in this paper, focusing on the higher circumferential vibrations of a thin cylindrical shell having freely supported end at the downstream side, the unstable phenomena of a thin cylindrical shell are clarified by using the root locus due to the complex eigenvalue analysis. The critical fluid velocities for the divergence and the flutter phenomena of a thin cylindrical shell subjected to an axial leakage flow are analyzed and defined. Moreover, comparing with the uncoupled vibration modes , the coupled vibration modes are shown to be considerably changed between before and after the unstable phenomenon. Finally, regarding the dimensions of a thin cylindrical shell, the velocity of an axial leakage flow, the characteristics of flowing fluid and the clearance of an annular gap as the parameters, the physical influences of parameters on the threshold of the instability of the coupled vibration are considered and discussed.

EQUATION OF MOTION

Let us consider the thin cylindrical shells subjected to the axial leakage flow shown in Figure 1. The inner shell is set to be elastic, on the other hand, the outer shell is set to be rigid, and both are set coaxially. On the other

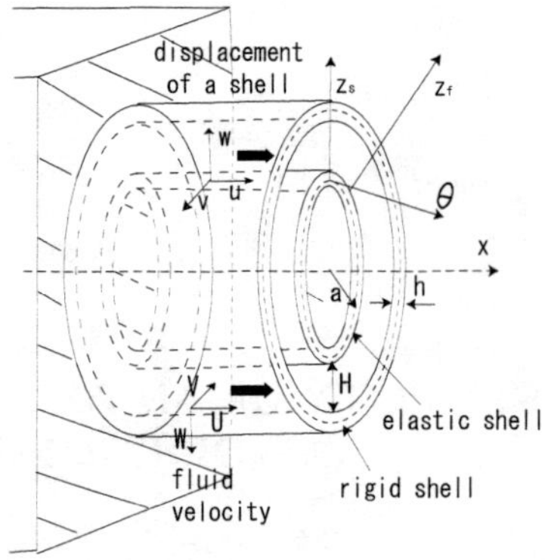

Figure 1. Analytical model of a thin cylindrical shell

hand, even if the inner structure is a rigid cylinder and the outer structure is an elastic shell, this theory can be similarly applied by changing the acting direction of the fluid pressure. The leakage flow passes through the annulus between two shells. The inner elastic shell vibrates due to the fluid force of the axial leakage flow. In this paper, the annular gap is considered to be sufficiently small compared with the radius and the length of a shell.

The equations of motion of an incompressible fluid based on the Navier-Stokes equations are given by equations of continuity and momentum:

$$\frac{\partial U}{\partial x}+\frac{\partial V}{\partial y}+\frac{\partial W}{\partial z_f}=0 \tag{1}$$

$$\frac{\partial U}{\partial t}+U\frac{\partial U}{\partial x}+V\frac{\partial U}{\partial y}+W\frac{\partial U}{\partial z_f}=-\frac{1}{\rho_f}\frac{\partial P}{\partial x}+\nu_f\frac{\partial^2 U}{\partial z_f^2} \tag{2}$$

$$\frac{\partial V}{\partial t}+U\frac{\partial V}{\partial x}+V\frac{\partial V}{\partial y}+W\frac{\partial V}{\partial z_f}=-\frac{1}{\rho_f}\frac{\partial P}{\partial y}+\nu_f\frac{\partial^2 V}{\partial z_f^2}. \tag{3}$$

The equations of a thin cylindrical shell are given by the Flügge's equations:

$$C\left\{\frac{\partial^2 u}{\partial x^2}+\frac{1-\nu_s}{2a^2}\frac{\partial^2 u}{\partial \theta^2}+\frac{1+\nu_s}{2a}\frac{\partial^2 v}{\partial x\partial\theta}+\frac{\nu_s}{a}\frac{\partial w}{\partial x}\right\}$$
$$+D\left\{\frac{1-\nu_s}{2a^4}\frac{\partial^2 u}{\partial\theta^2}-\frac{1}{a}\frac{\partial^3 w}{\partial x^3}+\frac{1-\nu_s}{2a^3}\frac{\partial^3 w}{\partial x\partial\theta^2}\right\}$$
$$-\mu\frac{\partial^2 u}{\partial t^2}+p_x=0 \tag{4}$$

$$C\left\{\frac{1}{a^2}\frac{\partial^2 v}{\partial\theta^2}+\frac{1-\nu_s}{2}\frac{\partial^2 v}{\partial x^2}+\frac{1+\nu_s}{2a}\frac{\partial^2 u}{\partial x\partial\theta}+\frac{1}{a^2}\frac{\partial w}{\partial\theta}\right\}$$
$$+D\left\{\frac{3(1-\nu_s)}{2a^2}\frac{\partial^2 v}{\partial x^2}-\frac{3-\nu_s}{2a^2}\frac{\partial^3 w}{\partial x^2\partial\theta}\right\}$$

$$-\mu\frac{\partial^2 v}{\partial t^2}+p_\theta-\frac{1}{a}\frac{\partial m_\theta}{\partial\theta}=0 \tag{5}$$

$$-C\frac{1}{a}\left\{\frac{1}{a}\frac{\partial v}{\partial\theta}+\frac{w}{a}+\nu_s\frac{\partial u}{\partial x}\right\}$$
$$+D\left\{\frac{1}{a}\frac{\partial^3 u}{\partial x^3}-\frac{1-\nu_s}{2a^3}\frac{\partial^3 u}{\partial x\partial\theta^2}+\frac{3-\nu_s}{2a^2}\frac{\partial^3 v}{\partial x^2\partial\theta}\right.$$
$$-\frac{\partial^4 w}{\partial x^4}+\frac{2}{a^2}\frac{\partial^4 w}{\partial x^2\partial\theta^2}+\frac{2}{a^4}\frac{\partial^2 w}{\partial\theta^2}$$
$$\left.+\frac{1}{a^4}\frac{\partial^4 w}{\partial\theta^4}+\frac{1}{a^4}w\right\}$$
$$-\mu\frac{\partial^2 w}{\partial t^2}+p_z-\frac{\partial m_x}{\partial x}-\frac{1}{a}\frac{\partial m_\theta}{\partial\theta}=0, \tag{6}$$

where the flexural rigidity of a shell C, D are $C=Eh/(1-\nu_s^2)$, $D=Eh^3/12(1-\nu_s^2)$.

FLOW RATE AND BOUNDARY CONDITIONS FOR FLOW

Now, introducing the axial and the circumferential flow rate $Q_x=\int_0^H U dz_f$, $Q_y=\int_0^H V dz_f$, where H is the gap, and integrating the equations (1)-(3), the flow velocity W disappears by using boundary conditions between a shell and a fluid and the following equations which describe the relation among the flow rate, the pressure and the gap are obtained.

$$\frac{\partial Q_x}{\partial x}+\frac{\partial Q_y}{\partial y}+\frac{\partial H}{\partial t}=0 \tag{7}$$

$$\frac{1}{\rho_f}\frac{\partial P}{\partial x}=-\frac{1}{H}\left\{\frac{\partial Q_x}{\partial t}+\frac{\partial}{\partial x}\left(\frac{Q_x^2}{H}\right)+\frac{\partial}{\partial y}\left(\frac{Q_xQ_y}{H}\right)+\frac{12\nu_f Q_x}{H^2}\right\} \tag{8}$$

$$\frac{1}{\rho_f}\frac{\partial P}{\partial y}=-\frac{1}{H}\left\{\frac{\partial Q_y}{\partial t}+\frac{\partial}{\partial x}\left(\frac{Q_xQ_y}{H}\right)+\frac{\partial}{\partial y}\left(\frac{Q_y^2}{H}\right)+\frac{12\nu_f Q_y}{H^2}\right\}. \tag{9}$$

The boundary conditions of the pressure at the inlet and outlet are obtained as follows:

$$P(0,y,t)=P_{in}-(1+\zeta_{in})\frac{\rho_f Q_x^2(0,y,t)}{2H^2(0,y,t)} \tag{10}$$

$$P(\ell,y,t)=P_{out}+\zeta_{out}\frac{\rho_f Q_x^2(\ell,y,t)}{2H^2(\ell,y,t)}, \tag{11}$$

where P_{in} and P_{out} are the pressures just in front of the inlet and just behind the outlet in an axial leakage flow.

UNSTEADY PRESSURE

The gap, the pressure and the flow rate can be divided into steady and unsteady parts as $H=\bar{H}+\Delta H(x,y,t)$, $P=\bar{P}(x)+\Delta P(x,y,t)$, $Q_x=\bar{Q}_x+\Delta Q_x(x,y,t)$, $Q_y=\Delta Q_y(x,y,t)$, where the bar and Δ imply the steady and unsteady values. The steady circumferential flow rate is neglected because it is considered to be smaller than the axial flow rate. Moreover, $\bar{H}$ and $\bar{Q}_y$ are constant. By applying the perturbation method and neglecting higher order terms of unsteady components, equations (7)-(11) can be linearized as follows:

$$\frac{\partial\Delta Q_x}{\partial x}+\frac{\partial\Delta Q_y}{\partial y}=-\frac{\partial\Delta H}{\partial t} \tag{12}$$

$$\frac{1}{\rho_f}\frac{\partial\Delta P}{\partial x}=-\frac{1}{\bar{H}}\frac{\partial\Delta Q_x}{\partial t}-\frac{\bar{Q}_x}{\bar{H}^2}\frac{\partial\Delta Q_x}{\partial x}-\frac{12\nu_f}{\bar{H}^3}\Delta Q_x+\frac{\bar{Q}_x^2}{\bar{H}^3}\frac{\partial\Delta H}{\partial x}+\frac{\bar{Q}_x}{\bar{H}^2}\frac{\partial\Delta H}{\partial t}+\frac{36\nu_f\bar{Q}_x}{\bar{H}^4}\Delta H \tag{13}$$

$$\frac{1}{\rho_f}\frac{\partial\Delta P}{\partial y}=-\frac{1}{\bar{H}}\frac{\partial\Delta Q_y}{\partial t}-\frac{\bar{Q}_x}{\bar{H}^2}\frac{\partial\Delta Q_y}{\partial x}-\frac{12\nu_f}{\bar{H}^3}\Delta Q_y \tag{14}$$

$$\Delta P(0,y,t)=-(1+\zeta_{in})\frac{\rho_f\bar{Q}_x^2}{\bar{H}^2}\times\left\{\frac{\Delta Q_x(0,y,t)}{\bar{Q}_x}-\frac{\Delta H(0,y,t)}{\bar{H}}\right\} \tag{15}$$

$$\Delta P(\ell,y,t)=\zeta_{out}\frac{\rho_f\bar{Q}_x^2}{\bar{H}^2}\left\{\frac{\Delta Q_x(\ell,y,t)}{\bar{Q}_x}-\frac{\Delta H(\ell,y,t)}{\bar{H}}\right\}. \tag{16}$$

Eliminating the flow rates from these equations, the relation between the unsteady pressure and the unsteady gap can be derived as follows:

$$\frac{1}{\rho_f}\left(\frac{\partial^2}{\partial x^2}+\frac{\partial^2}{\partial y^2}\right)\Delta P=\frac{1}{\bar{H}}\frac{\partial^2\Delta H}{\partial t^2}+\frac{2\bar{Q}_x}{\bar{H}^2}\frac{\partial^2\Delta H}{\partial x\partial t}+\frac{12\nu_f}{\bar{H}^3}\frac{\partial\Delta H}{\partial t}+\frac{\bar{Q}_x^2}{\bar{H}^3}\frac{\partial^2\Delta H}{\partial x^2}+\frac{36\nu_f\bar{Q}_x}{\bar{H}^4}\frac{\partial\Delta H}{\partial x}. \tag{17}$$

MODE FUNCTION

In this paper, focusing on the circumferential vibrations, the fluid-structure coupling and the unstable behavior are considered and discussed.

The displacements of a shell subjected to an axial leakage flow are assumed to be approximated by superposing the eigenmode functions in a vacuum. Employing the eigenmodes in the axial, the circumferential and the radial directions, $\hat{u}_{mn}(x)$, $\hat{v}_{mn}(x)$ and $\hat{w}_{mn}(x)$, respectively, the fluid-coupled displacements $u(x,\theta,t)$, $v(x,\theta,t)$ and $w(x,\theta,t)$ are expanded as follows:

$$\begin{pmatrix} u(x,\theta,t) \\ v(x,\theta,t) \\ w(x,\theta,t) \end{pmatrix} = \sum_{m=1}^{\infty}\sum_{n=0}^{\infty} \begin{pmatrix} \hat{u}_{mn}(x)\cos n\theta \\ \hat{v}_{mn}(x)\sin n\theta \\ \hat{w}_{mn}(x)\cos n\theta \end{pmatrix} \xi_{mn}(t), \quad (18)$$

where $\xi_{mn}(t)$ is the function of time. $\hat{u}_{mn}(x)$, $\hat{v}_{mn}(x)$ and $\hat{w}_{mn}(x)$ are obtained by the exact solution.

Solution Satisfying Flügge's Equation

The solution which satisfies equations(4)-(6) is assumed to be expressed in the circumferential direction term, axial direction one and time function separately as follows[(12),(13)]:

$$\left.\begin{aligned} u &= U'(x)\cos(n\theta)(A\cos\omega t + B\sin\omega t) \\ v &= V'(x)\sin(n\theta)(A\cos\omega t + B\sin\omega t) \\ w &= W'(x)\cos(n\theta)(A\cos\omega t + B\sin\omega t) \end{aligned}\right\}, \quad (19)$$

where ω is eigenfrequency in a vacuum. Substituting equation(19) into equations(4)-(6) and then as U', V', W' in equations(19) are the required mode functions, let consider U', V', W' to be expressed as $U' = c_1e^{\lambda x}$, $V' = c_2e^{\lambda x}$, $W' = c_3e^{\lambda x}$. Finaly, the following equation is obtained:

$$e^{\lambda x}\begin{bmatrix} U_1' & U_2' & U_3' \\ V_1' & V_2' & V_3' \\ W_1' & W_2' & W_3' \end{bmatrix}\begin{bmatrix} c_1 \\ c_2 \\ c_3 \end{bmatrix} = \vec{O}, \quad (20)$$

where $U_1' \sim U_3', V_1' \sim V_3', W_1' \sim W_3'$ are functions of ω and λ. The conditions of free vibration in equation(20) is that c_1, c_2, c_3 must have not zero solutions. Therefore,

$$\det\begin{bmatrix} U_1' & U_2' & U_3' \\ V_1' & V_2' & V_3' \\ W_1' & W_2' & W_3' \end{bmatrix} = a_0\lambda^8 + a_1\lambda^6 + a_2\lambda^4 + a_3\lambda^2 + a_4 = 0 \quad (21)$$

has to be satisfied. Here, $a_0 \sim a_4$ are functions of ω. Solving equation(21), $\lambda_j (j = 1\cdots, 8)$ are obtained, then the general solutions concerning U', V', W' are given as follows:

$$\left.\begin{aligned} U'(x) &= \textstyle\sum_{j=1}^{8} c_3{}^{(j)}k_u(\lambda_j)\,e^{\lambda_j x} \\ V'(x) &= \textstyle\sum_{j=1}^{8} c_3{}^{(j)}k_v(\lambda_j)\,e^{\lambda_j x} \\ W'(x) &= \textstyle\sum_{j=1}^{8} c_3{}^{(j)}e^{\lambda_j x} \end{aligned}\right\}, \quad (22)$$

where $k_u(\lambda_j)$ and $k_v(\lambda_j)$ are functions of $c_1/c_3, c_2/c_3$ obtained by solving equation(20).

Boundary Condition of a Shell

When a thin cylindrical shell has the boundary conditions in which one end is clamped and the other is free, 8 equations of condition can be given. That is, substituting equation(22) into these equations of boundary conditions, the following equation is obtained:

$$Q\left\{c_3{}^{(1)}\cdots c_3{}^{(8)}\right\}^T = 0, \quad (23)$$

where Q is the matrix having 8×8 elements which are functions of ω and λ. The solutions of λ in equation(21) are complex solutions consisting of complex conjugates. Therefore, the following vector $\vec{d}$ consisting of 16 elements is introduced.

$$\vec{d} = [Re(c_3^{(1)})\ldots Re(c_3^{(8)})\ Im(c_3^{(1)})\ldots Im(c_3^{(8)})]^T. \quad (24)$$

Equation(23) is modified by using $\vec{d}$ as follows:

$$\Big[Re(Q) \vdots -Im(Q)\Big]\vec{d} = \vec{0}. \quad (25)$$

Equation(25) is expressed by a matrix of 8 × 16. Besides, $[c_3^{(1)}\ldots c_3^{(8)}]^T$ in equation(23) are complex conjugates because λ in equation(23) have complex conjugates. Then, as the real parts are equal and the imaginary parts have equal absolute values but not equal, the 8 following equations are given:

$$[R]\vec{d} = \vec{0} \quad (26)$$

Combining equations(25) and (26) together, 16 equations are obtained as follows:

$$[S]\vec{d} = \vec{O}, \quad (27)$$

where

$$S = \begin{bmatrix} Re(Q) \vdots -Im(Q) \\ R \end{bmatrix}. \quad (28)$$

For the condition that $\vec{d}$ have nontrivial solutions, the following equation must be satisfied:

$$\det[S] = 0. \quad (29)$$

Solving the simultaneous equations consisting of equations (21) and (29), ω and λ can be obtained. Furthermore, substituting ω and λ into eqnuation (27), vector $\vec{d}$ can be given by getting the nontrivial solutions. Following the above-mentioned procedure, ω_{mn} and λ_{mn} in the m-th axial mode and the n-th circumferential mode can be obtained. When $A_{mnj} = d_j + \mathbf{i}d_{j+8}$ ($\mathbf{i}$ is $\sqrt{-1}$) using vectors $\vec{d}$ determined by ω_{mn} and λ_{mn} is set, the eigenmode functions in the m-th axial mode and the n-th circumferential mode are expressed finally as follows:

$$\left.\begin{array}{ll} \hat{u}_{mn}(x) = \sum_{j=1}^{8} A_{mnj} k_u(\lambda_{mnj}) e^{\lambda_{mnj}x} \\ \hat{v}_{mn}(x) = \sum_{j=1}^{8} A_{mnj} k_v(\lambda_{mnj}) e^{\lambda_{mnj}x} \\ \hat{w}_{mn}(x) = \sum_{j=1}^{8} A_{mnj} e^{\lambda_{mnj}x} \end{array}\right\}. \quad (30)$$

COUPLED FLUID FORCE

Here, in this section, the unsteady pressure ΔP is derived and fluid forces and moments are calculated more specifically. Since the unsteady gap width ΔH coincides with the radial displacement of a shell, the following relation can be obtained: $\Delta H = w$. Using this relation and equation(18), equation(17) is rewritten as follows:

$$\frac{1}{\rho_f}\left(\frac{\partial^2}{\partial x^2} + \frac{\partial^2}{\partial y^2}\right)\Delta P(x,\theta,t) = \sum_{m=1}^{\infty}\sum_{n=0}^{\infty}\cos n\theta \Big\{ A_{mn}(x)\ddot{\xi}_{mn}(t) + B_{mn}(x)\dot{\xi}_{mn}(t) + C_{mn}(x)\xi_{mn}(t) \Big\}, \quad (31)$$

where the coefficients $A_{mn}(x)$, $B_{mn}(x)$, $C_{mn}(x)$ are the functions given by the modal function $\hat{w}_{mn}(x)$ and its derivatives. Since y is described as $y{=}a\theta$, the left hand side of equation(31) is the function of x, θ and t. The unsteady pressure ΔP is assumed to be written as $\Delta P(x,\theta,t) = \sum_{n=0}^{\infty} \Delta P_n(x,t)\cos n\theta$, where ΔP_n means the unsteady pressure at $\theta = 0$. Substituting this into equation(31), the ordinary differential equation is obtained. Solving the obtained ordinary differential equation under the linearlized boundary conditions based on equations(10) and (11), the relation between the unsteady pressure $\Delta P(x,\theta,t)$ and the modal displacement is obtained:

$$\Delta P(x,\theta,t) = \sum_{m=1}^{\infty}\sum_{n=0}^{\infty}\cos n\theta \Big\{ \bar{m}_{mn}(x)\ddot{\xi}_{mn}(t) + \bar{c}_{mn}(x)\dot{\xi}_{mn}(t) + \bar{k}_{mn}(x)\xi_{mn}(t) \Big\}. \quad (32)$$

The functions $\bar{m}_{mn}(x)$, $\bar{c}_{mn}(x)$ and $\bar{k}_{mn}(x)$ are coefficient functions. Their concrete forms are omitted.

FLUID-SHELL COUPLED EQUATION AND STABILITY ANALYSIS

The fluid forces acting on a shell are given by $p_x = -\mu_f \partial\Delta U/\partial z_f|_{z_f=0}$, $m_x = \mu_f \partial\Delta U/\partial z_f|_{z_f=0}(h/2)$, $p_\theta = -\mu_f \partial\Delta V/\partial z_f|_{z_f=0}$, $m_\theta = \mu_f \partial\Delta V/\partial z_f|_{z_f=0}(h/2)$, $p_z = \Delta P(x,\theta,t)|_{z_f=0} = \Delta P(x,\theta,t)$. Assuming the velocity distribution as $\Delta U = (1/2\mu_f)\partial\Delta P/\partial x z_f(z_f - \bar{H})$, $\Delta V = (1/2\mu_f)\partial\Delta P/\partial\theta z_f(z_f - \bar{H})$, the fluid forces become $p_x = -(\bar{H}/2)\partial\Delta P/\partial x$, $p_\theta = (\bar{H}/2)\partial\Delta P/\partial\theta$, $p_z = \Delta P(x,\theta,t)$, $m_x = (h\bar{H}/4)\partial\Delta P/\partial x$, $m_\theta = (h\bar{H}/4)\partial\Delta P/\partial\theta$, where $\mu_f = \rho_f\nu_f$.

Substituting these and the equation(18) into the equations of a thin cylindrical shell (4)-(6) and utilizing the orthogonality of the mode of the axial and the circumferential directions, the following fluid-shell coupled equation of motion is obtained, that is, multiplying equations(4)-(6) by the eigenfunction of the p-th in the axial direction and the q-th in the circumferential direction and integrating the resultant equation along the axial direction (0 to ℓ) and the circumferential direction (0 to 2π):

$$\ddot{\xi}_{pq}(t) + \omega_{pq}^2\xi_{pq}(t) = -\sum_{m=1}^{\infty}\Big\{ \hat{m}_{pmq}\ddot{\xi}_{mq}(t) + \hat{c}_{pmq}\dot{\xi}_{mq}(t) + \hat{k}_{pmq}\xi_{mq}(t) \Big\}, \quad (33)$$

where ω_{pq} is the eigenfrequency of the shell in a vacuum. The left hand side of the above equation is described on only the shell, therefore it can be understood that the left hand side is orthogonalized with the p-th axial mode and the q-th circumferential mode. On the other hand, the fluid forces in the right hand side in equation(33) can not be orthogonalized. That is, the combination of the function of time with $m = 1, 2, \cdots, \infty$ is left. By this equation, even in the fluid-shell coupled vibration subjected to the axial leakage flow, the vibration modes in the axial direction and the circumferential one can be transformed to the product of the function of each variable when a coupled eigenvalue analysis is performed for equation(33) and the orthogonality condition among the coupled modes can be applicable. Therefore, it is seen that similarly to the modes in a vacuum, the axisymmetric, the lateral and the circumferential vibrations are considered separately even in a fluid-shell coupling. Truncating the number of mode as M, employing vectors and matrices form, and changing q to n, equation(33) can be rewritten as follows:

$$[M_{s,n} + M_{a,n}]\{\ddot{\xi}_n(t)\} + [C_{a,n}]\{\dot{\xi}_n(t)\} + [K_{s,n} + K_{a,n}]\{\xi_n(t)\} = \{0\}, \quad (34)$$

where $\{\xi_n(t)\} = \{\xi_{1n}(t), \xi_{2n}(t), \cdots, \xi_{Mn}(t)\}^T$,

$[M_{s,n}] = I_{M\times M}$, $[K_{s,n}] = diag\{\omega_{1n}^2, \omega_{2n}^2, \cdots, \omega_{Mn}^2\}$, $[M_{a,n}] = [\hat{m}_{mpn}]_{1\le m,p\le M}$, $[C_{a,n}] = [\hat{c}_{mpn}]_{1\le m,p\le M}$, $[K_{a,n}] = [\hat{k}_{mpn}]_{1\le m,p\le M}$. n=0, n=1 and n ≥2 denote the axisymmetric, the lateral and the circumferential vibrations respectively. By calculating the complex eigenvalue s and plotting the root locus, the stability of a system can be evaluated.

Here, there are $2M$ complex eigenmodes of equation(34) for fixed n. M pairs of eigenvalues exist. Each pair is a set of complex conjugate numbers or a pair of two real numbers. Let us call the pairs the $1, 2, 3, \cdots M$-th eigenvalues. When the imaginary part of a complex eigenvalue is not zero, the complex eigenvalue with a positive imaginary part in a pair of complex conjugate eigenvalues is adopted. On the other hand, when the imaginary part is zero, the eigenvalue with a larger real part in a pair of eigenvalues is adopted. Thus, the focused M complex eigenvalues are written as $\{\lambda_{1,n}, \lambda_{2,n}, \cdots, \lambda_{M,n}\}$. Let the complex eigenvectors corresponding to $\lambda_{k,n}$ be $\{\Psi_{k,n}\}$, and the m-th element of $\{\Psi_{k,n}\}$ be $\Psi_{k,n,m}$. Moreover, let $\eta_{k,n}(t)$ be the coupled function of time corresponding to $\{\Psi_{k,n}\}$. Therefore, the coupled k-th complex eigenmodes are expressed as follows:

$$\left.\begin{aligned} u_{k,n}^c(x,\theta) &= \sum_{m=1}^{M} \hat{u}_{mn}(x)\cos(n\theta)\Psi_{k,n,m} \\ v_{k,n}^c(x,\theta) &= \sum_{m=1}^{M} \hat{v}_{mn}(x)\sin(n\theta)\Psi_{k,n,m} \\ w_{k,n}^c(x,\theta) &= \sum_{m=1}^{M} \hat{w}_{mn}(x)\cos(n\theta)\Psi_{k,n,m} \end{aligned}\right\} \quad (35)$$

Here, note the coupled complex eigenvalues and eigenmodes vary when the flow velocity varies. By employing these equations, the displacement of a cylindrical shell can be expressed as follows:

$$\begin{pmatrix} u(x,\theta,t) \\ v(x,\theta,t) \\ w(x,\theta,t) \end{pmatrix} = \sum_{k=1}^{M}\sum_{n=0}^{N} \begin{pmatrix} u_{k,n}^c(x,\theta) \\ v_{k,n}^c(x,\theta) \\ w_{k,n}^c(x,\theta) \end{pmatrix} \eta_{k,n}(t). \quad (36)$$

The fact that the imaginary part of complex eigenvalue s satisfies $Im(s) \neq 0$ and the real part $Re(s) > 0$ implies dynamically unstable (flutter) and $Im(s) = 0$ and $Re(s) > 0$ implies statically unstable (divergence). According to these procedures, the critical flow velocity and complex eigenmode shape for the leakage-flow induced vibration can be evaluated.

NUMERICAL ANALYSIS

Dimensions for Numerical Analysis Studies

As numerical analysis examples, let us consider the inner shell is elastic, and the outer shell is rigid as shown in Figure 1. Three types of shell models are considered here: The first model is the shorter elastic shell, the second model is the longer one and the third model is the thiner one comparing with the first model shown in Table 1. The elastic shell is assumed to be a cantilever shell stucture in which the upperstream side is fixed and the downstream side is free. The fluid is water or air. The elastic shell is considered to be made of steel. Hereafter, let us call these models as Case 1, Case 2 and Case 3.

Table 1. Dimensions of numerical models

	Case 1	Case 2	Case 3
length [m]	0.5	3	0.5
radius [m]	1	1	1
thickness [m]	0.002	0.002	0.0002
gap width [m]	0.003	0.003	0.003
ℓ/a	0.5	3	0.5
a/h	500	500	5000

Eigenvalue Analysis of Numerical Models

Figures 2 and 3 show the eigenfrequencies in dry and wet conditions for Case 1 and Case 2 respectively. Their

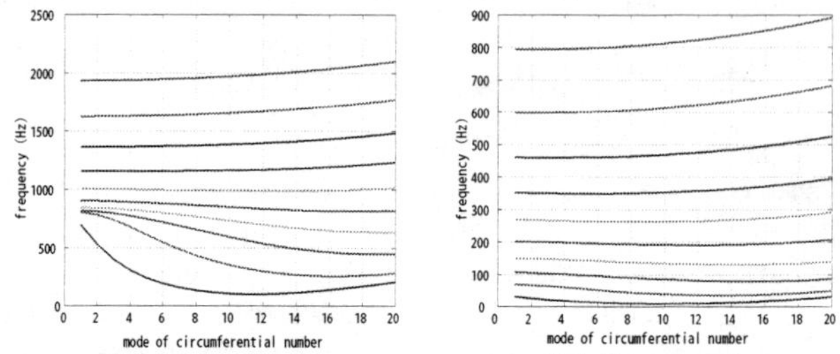

(a) Dry mode (b) Wet mode

Figure 2. Eigenvalue analysis in Case 1

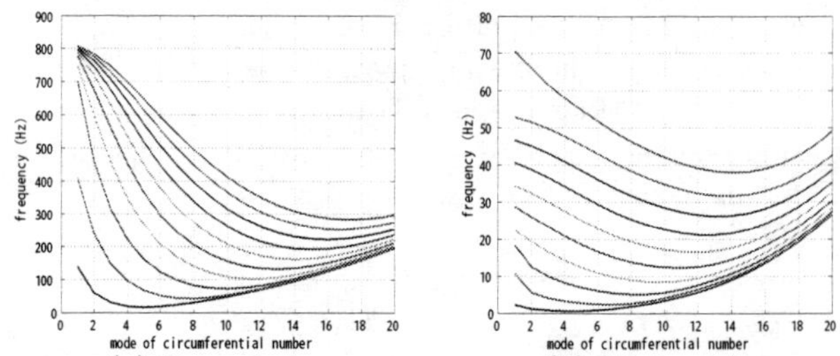

(a) Dry mode (b) Wet mode

Figure 3. Eigenvalue analysis in Case 2

curves show the eigenfrequencies of the m-th axial mode from lower number $m = 1, 2, \cdots 10$ in order. The dry condition means the eigenvalues in a vacuum, and the wet means those in a still fluid. Hereafter let us call them dry modes and wet modes respectively. In these figures, the axial mode number m is taken as a parameter. As a fluid is water, the eigenfrequencies of wet modes become much lower due to the added mass effect of the fluid, especially for a small annulus gap. In Case 1, the eigenfrequencies of dry modes show the minimum at the mode $(m, n) = (1, 11)$. On the other hand, those of wet modes show that at the mode $(m, n) = (1, 10)$. In Case 2, the eigenfrequencies of dry modes show the minimum at the mode $(m, n) = (1, 5)$. On the other hand, those of wet modes show that at the mode $(m, n) = (1, 4)$.

Stability on Circumferential Vibrations

Concerning the circumferential vibrations of a thin cylindrical shell subjected to an axial leakage flow, their dynamic stability is investigated in this subsection.

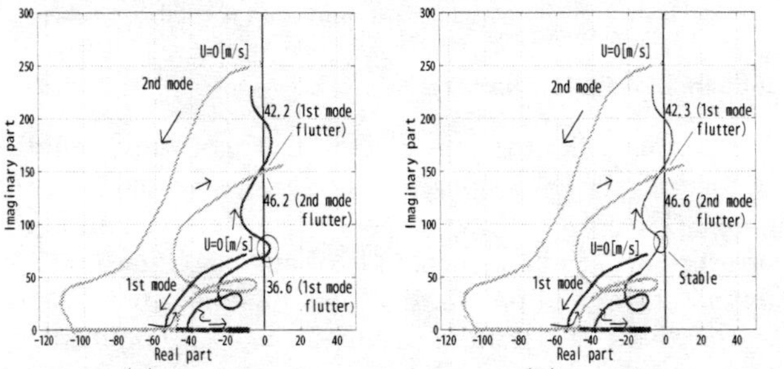

(a) n=10 (b) n=11

Figure 4. Root locus in Case 1

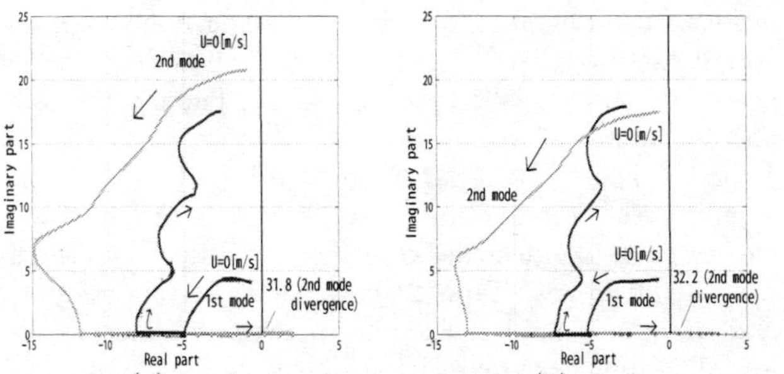

(a) n=4 (b) n=5

Figure 5. Root locus in Case 2

Figures 4 and 5 show the root loci for Case 1 ($n = 10, 11$) and Case 2 ($n = 4, 5$) respectively. Regarding the 10th circumferential vibration in Case 1, the 1st axial mode goes into flutter phenomenon at $U = 36.6m/s$ and the 2nd mode goes into flutter phenomenon at $U = 46.2m/s$. For the 11th circumferential vibration in Case 1, the 1st axial mode goes into flutter phenomenon at $U = 42.3m/s$ and the 2nd mode goes into flutter phenomenon at $U = 46.6m/s$. When both results are compared, the shapes of both root locus curves are found to be similar. However, forcusing on the curve of the root locus in the first unstable region of the 1st axial mode in Figure 4 (a), the same curve of the root locus in Figure 4 (b) is stable. Therefore, the curve of the root locus in figure 4 (b) can be seen to be shifted to the left hand side comparing with Figure 4 (a). Moreover increasing the fluid velocity, the 1st and the 2nd axial modes become stable after they became unstable. Then these axial modes become unstable again. Regarding the 4th circumferential vibration in Case 2, the 2nd axial mode goes into divergence phenomenon at $U = 31.8m/s$. For the 5th circumferential vibration in Case 2, the 2nd axial mode goes into divergence phenomenon at $U = 32.2m/s$. Figure 5 (a) and (b) are quite similar. The 1st axial modes are stable and their eigenfrequencies increase increasing the fluid velocity .

Complex Eigenmode

In this subsection, comparing with the uncoupled vibration modes, the coupled vibration modes are shown to have changed considerably between before and after the unstable phenomenon.

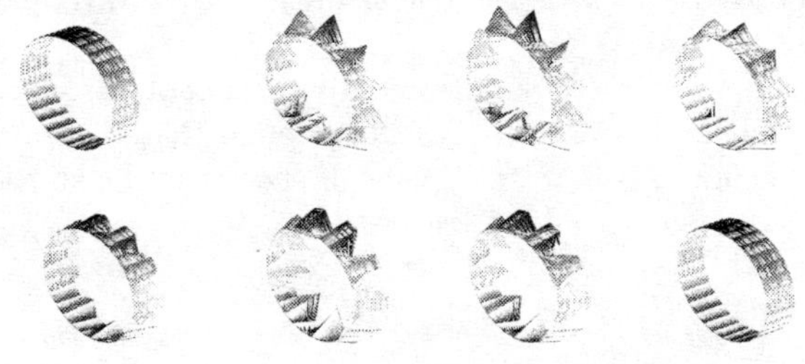

(a) $U = 0m/s$, 1st mode

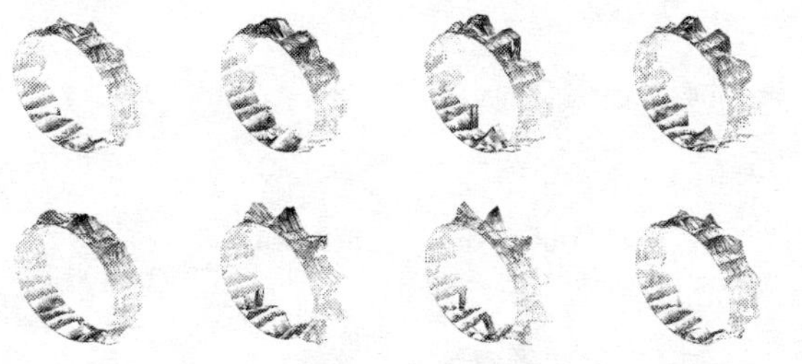

(b) $U = 36.6m/s$, flutter, 1st mode

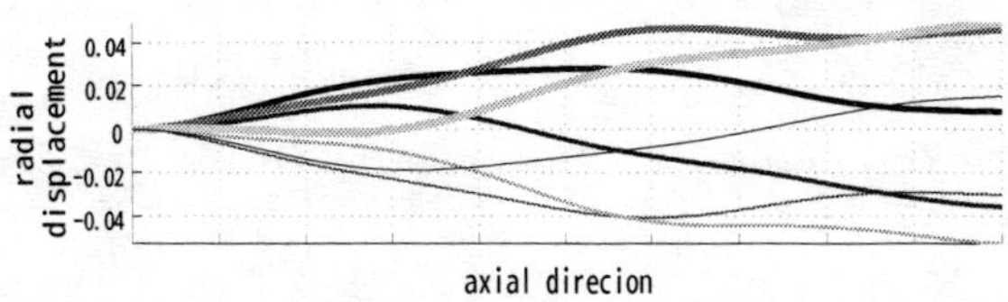

(c) $U = 36.6m/s$, flutter, 1st mode, radial displacement

Figure 6. Complex eigenmode of the circumferential vibration (Case 1, 1st mode, n=10)

Figures 6 and 7 show the mode shapes at every instance, $t = 0, T/7, \cdots, 6T/7, T$ over one period T. Figure 6 shows the 1st complex eigenmode of the circumferential vibration in Case 1 which becomes unstable. Figure 6 (a) indicates that in the still fluid at $U = 0m/s$, Figure 6 (b) indicates the flutter mode at $U = 36.6m/s$, and Figure 6 (c) indicates only the radial component of the flutter mode at $U = 36.6m/s$ in order to investigate the generation of a traveling wave. The complex eigenmode at $U = 0m/s$ is almost the same as the dry mode. However, the flutter mode shape at $U = 36.6m/s$ is fairly different from at $U = 0m/s$. Moreover, the location of the maximum amplitude of the flutter mode over one period is seen not to be fixed axially and to move along the axial direction as shown in Figure 6 (c) clearly. This behavior can be considered to mean the generation of a traveling wave axially. Figure 7 shows the 2nd complex eigenmode in Case 2 which goes into the divergence phenomenon. Figure 7 (a) indicates the 2nd eigenmode in the still fluid at $U = 0m/s$ and Figure 7 (b) indicates the divergence at $U = 31.8m/s$. The right hand side of Figure 7 (b) is the one side-view of the divergence mode.

Comparing the circumferential vibration of the shorter shell and the longer shell, it can be found that the flutter phenomenon is apt to take place in shorter shells and the divergence is apt to do in longer shells.

(a) $U = 0m/s$, 2nd mode

(b) $U = 31.8m/s$, divergence, 2nd mode

Figure 7. Complex eigenmode of the circumferential vibration (Case2, 2nd mode, n=4)

Influence of Gap Width

In this following, the influence of the gap width on the stability is examined. Figure 8 (a) shows the influence of the gap width on the critical velocity of the flutter phenomenon of the 10th circumferential vibration in Case 1. Figure 8 (b) shows the influence of the gap width on the critical velocity of the divergence phenomenon of the 4th circumferential vibration in Case 2. The small gap width can be seen to make vibration of a thin cylindrical shell subjected to an axial leakage flow more unstable.

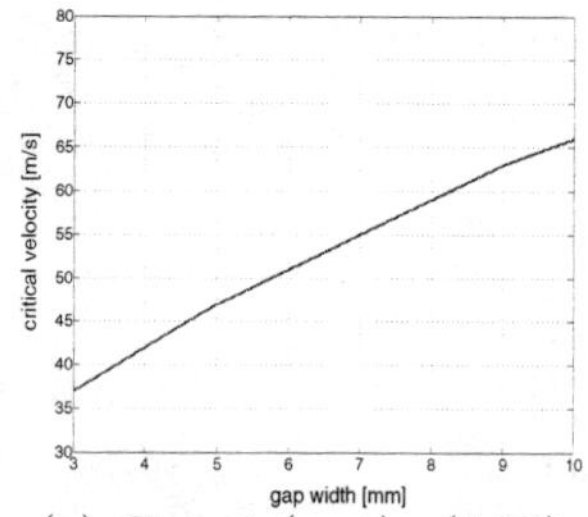

(a) Case 1,(m,n)=(1,10) flutter

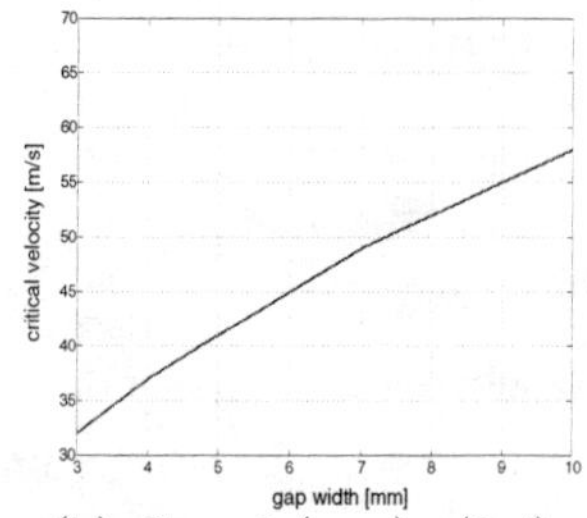

(b) Case 2,(m,n)=(2,4) divergence

Figure 8. Influence of the gap width on the critidal velocity

Influence of Fluid Characteristics

In this following, the influence of not only liquid described in above sections but also gas as a fluid is investigated. Except for water at room temperature and normal pressure as a liquid, four kinds of air which have different temperatures and pressures are adopted as a gas in numerical analysis. As fluid is assumed to be incompressible in this proposed analysis, the accuracy of numelical solutions can be considered to be hold approximately in low Mach number even if air as a fluid is adopted in numerical calculations. The dimensions of a shell having freely supported end at the downstream side utilized in numerical studies is Case 3 as already shown in Table 1. Though the length and the radius of a shell in Case 3 is the same as in Case 1, its thickness in Case 3 is one tenth in Case 1 in order to minimize the reduction of the fluid-structure interaction effect due to the decrease of density of a fluid by changing from liquid to gas. Table 2 shows the characteristics of air and water as a fluid utilized in numerical parametric studies. The dimensions are pressure: $P[MPa]$, temperature: $T'[K]$, sound velocity: $c[m/s]$, density of a fluid: $\rho_f[kg/m^3]$ and kinematic viscosity: $\nu_f[m^2/s]$.

Figure 9 shows eigenfrequencies in dry and wet conditions for the representative numerical calculation cases in Table 2. Figure 9 (a) shows those in the dry condition of Case 3 in Table 1. Besides, these frequencies are quite similar those for the wet condition in still air with room temperature and normal pressure ($T' = 300K, P = 0.1MPa$), and with high temperature and normal pressure ($T' = 900K, P = 0.1MPa$) when the annulus gap is quite large, and these are a little different when the annulus gap is small. Figure 9 (b) and (c) show those for wet condition in still air with room temperature and high pressure ($T' = 300K, P = 2MPa$) and with high temperature and high pressure ($T' = 900K, P = 2MPa$), respectively corresponding to Case 3 (c) and (d) in Table 2. Moreover, Figure 9 (d) shows those for wet condition in still water corresponding to Case 3 (e) in Table 2.

Table 2. Characteristics of fluid

	fluid	P	T'	c	ρ_f	ν_f
(a)	air	0.1	300	347	1.1609	1.604×10^{-5}
(b)	air	0.1	900	602	0.3867	1.04×10^{-4}
(c)	air	2	300	346	23.332	8.04×10^{-7}
(d)	air	2	900	604	7.6835	5.238×10^{-6}
(e)	water	0.1	300	/	998.2	1.004×10^{-6}

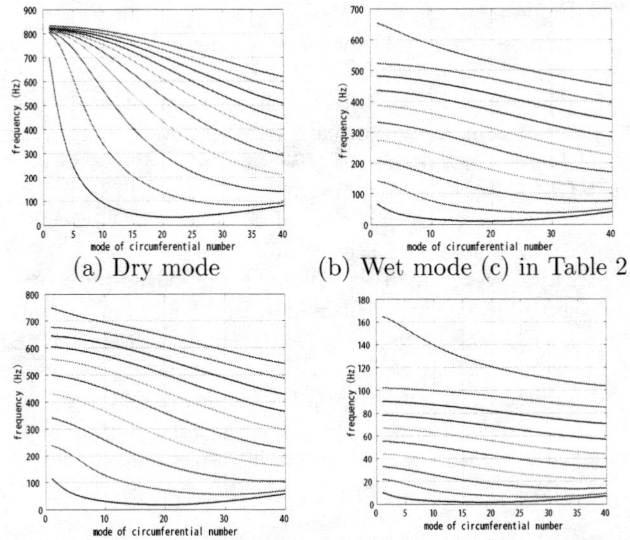

(a) Dry mode (b) Wet mode (c) in Table 2

(c) Wet mode (d) in Table 2 (d) Wet mode (e) in Table 2

Figure 9. Eigenvalue analysis in Case 3

Figure 10 shows the root loci for Case 3 ($n = 13$) corresponding to Table 2. Table 3 summarizes the critical velocities at which the vibration behaviors go into divergence or flutter phenomenon when the flow velocity increases. The dimensions are critical velocity: $U_{cr}[m/s]$ and Mach number: M'. Comparing Figure 10 (a) with (b), and Figure 10 (c) with (d), the critical velocity is found not to be changed so remarkably in spite of the reduction of fluid-structure interaction effect by little decrease of fluid density, and the damping effect by increase of the kinematic viscosity when the temperature of air becomes high, though the kinematic viscosity of water decreases when the temperature increases. On the other hand, comparing Figure 10 (a) with (c), and Figure 10 (b) with (d), the critical velocity is found to become much lower due to much increase of density and decrease of the kinematic viscosity when the pressure of air becomes high. Generally speaking, the influence of increasing pressure can be said to be greater than that of temperature for the selected dimension-range in the parameter studies as shown in Table 2. Moreover, comparing Figure 10 (e) with (a), the critical velocity is found to become much higher due to decrease of fluid density, and increase of the kinematic viscosity, when water as a fluid is changed to air.

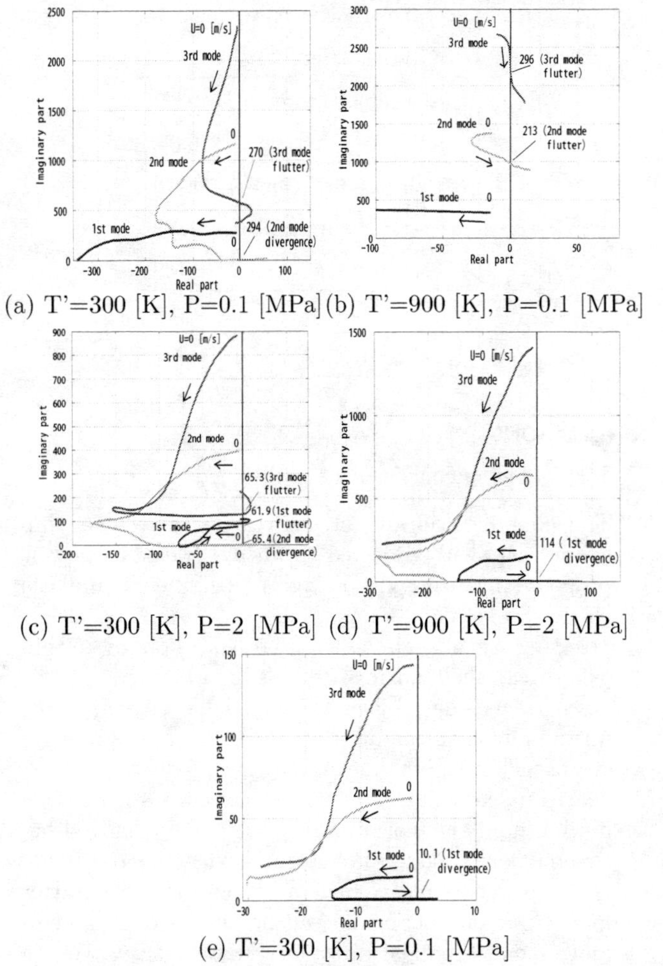

(a) T'=300 [K], P=0.1 [MPa] (b) T'=900 [K], P=0.1 [MPa]

(c) T'=300 [K], P=2 [MPa] (d) T'=900 [K], P=2 [MPa]

(e) T'=300 [K], P=0.1 [MPa]

Figure 10. Influence of a fluid

Thus, as the fluid density become much lower when a fluid is changed from water to gas, it can be generally said that the effect of fluid-structure interaction is reduced, and thus the critical velocity becomes higher, that is, the vibration behavior becomes stable. On the other hand, gas is generally apt to be utilized in high temperature and high pressure when it is used as a thermal medium in industrial plants such as in gas turbine facilities and so forth.

Therefore, comparing Figure 10 (d) in high temperature and high pressure air ($T' = 900K, P = 2MPa$) with Figure 10 (a) in room temperature and normal pressure air ($T' = 300K, P = 0.1MPa$), it can be understood that the vibration behavior becomes unstable when gas is utilized in high temperature and high pressure condition.

Table 3. Summary of critical velocity and Mach number

	U_{cr}	M'
(a)	274 (m=2, divergence)	0.85
	270 (m=3, flutter)	0.78
(b)	296 (m=2, flutter)	0.47
	213 (m=3, flutter)	0.35
(c)	61.9 (m=1, flutter)	0.179
	65.4 (m=2, divergence)	0.189
	65.3 (m=3, flutter)	0.189
(d)	114 (m=1, divergence)	0.189
(e)	10.1 (m=1, divergence)	/

CONCLUSIONS

The following conclusions are obtained:

(1) The coupling vibrational analysis method of a thin cylindrical shell subjected to an axial leakage flow is proposed by using the Flügge's shell theory and the Navier-Stokes equation.

(2) Focusing on the circumferential vibration of a thin cylindrical shell having freely supported end at the downstream side subjected to an axial leakage flow, the unstable phenomena are clarified by using the root locus based on the complex eigenvalue analysis.

(3) The coupled vibration modes between a shell and a flowing fluid are shown to be considerably changed between before and after the unstable phenomenon. Especially the location of maximum amplitude of the flutter mode over one period is seen not to be fixed and to move along the axial direction. A traveling wave can be found to generate axially.

(4) Comparing the circumferential vibration of the shorter shell and the longer shell, it can be found that the flutter phenomenon is apt to take place in shorter shells and the divergence is apt to do in longer shells.

(5) The smaller gap width is found to make a thin cylindrical shell subjected to an axial leakage flow more unstable.

(6) When a fluid is changed from water to air, the critical velocity is found to be higher due to the density of air and the vibration behavior becomes stable.

(7) It can be considered that the vibration behavior becomes unstable when gas is utilized in high temperature and high pressure condition.

REFERENCES

(1) Alford, J.S., (1964), Protection of labyrinth seals from flexural vibration, Journal of Engineering for Power, 141-148.

(2) Alford, J.S., (1967), Protecting turbomachinery from unstable and oscillatory flows, Journal of Engineering for Power, 513-528.

(3) Ziada, S., & Bühlman, E. T., Bolleter, U. (1987), Model tests on shell flutter due to a leakage flow, *Int. Conf. on Flow-Induced Vibrations Bowness-on Windermere*, England, M2, 521-529.

(4) Hobson, D. E. (1982), Fluid-elastic instabilities caused by flow in an annulus, *Proceedings of BNES 3rd Int. Conf. in Vibration in Nuclear Plant*, 440-461.

(5) Païdoussis, M. P. (1966), Dynamics of flexible slender cylinders in axial flow, part 1. theory, part 2. experiment, *Journal of Fluid Mechanics, 26*, 717-751.

(6) Païdoussis, M. P., et. al. (1984), Dynamics and stability of coaxial cylindrical shell containing flowing fluid, *Journal of Sound and Vibration, 97, 2*, 201-235.

(7) Chen, S. S., & Wambsganss, M. W. (1972), Parallel-Flow-Induced Vibration of Fuel Rods, *Nuclear Eng. and Design, 18*, 253-278.

(8) Fujita, K., Ito, T. & Baba, K. (1993), Study on the unstable vibration of cylindrical shells due to leakage flow, *Proc. of the Int. Conf. on Aero-Hydroelasticity*, 286-295.

(9) Fujita, K., & Shintani,A, & Ono, M, (2003), Axial leakage flow-induced vibration of a thin cylindrical shell with respect to axisymmetric vibration, Trans. of ASMi, J. of Pressure Vessel Thechnology, 125-2, 151-157.

(10) Fujita, K., & Shintani,A, & Ono, M,(2003), Axial leakage flow-induced vibration of a thin cylindrical shell with respect to lateral vibration, Trans. of ASME, J. of Pressure Vessel Thechnology, 125-2, 158-164.

(11) Fujita, K., & Shintani, A., (2003), Unstable Phenomena of a Thin Cylindrical Shell Subjected to Axial Leakage Flow, IUTAM Symposium on Integrated Modeling of Fully Coupled Fluid Structure Interactions Using Analysis, Computations and Experiments, Kluwer Aca. Pub., 331-349.

(12) Weingarten,V.I., (1964), Free vibration of thin cylindrical shells, AIAA Journal, VOL.2 No.4, 717-722.

(13) Fujita, K., & Saito, A, (2003), Free vibration and seismic analysis of a liquid storage thin cylindrical shell with unaxismmetric attached mass and stiffness, Proc. of ASME, PVP-Vol.466, 243-252.

Proceedings of IMECE04
2004 ASME International Mechanical Engineering Congress and Exposition
November 13-20, 2004, Anaheim, California USA

IMECE2004-59494

UNSTABLE VIBRATION OF SIMPLY SUPPORTED THIN CYLINDRICAL SHELLS SUBJECTED TO AXIAL LEAKAGE FLOWS USING FLÜGGE'S SHELL THEORY

Katsuhisa Fujita and Makoto Kato
Mechanical Systems Engineering, Graduate School of Engineering,
Osaka Prefecture University, 1-1, Gakuen-cho, Sakai-City, Osaka Prefecture, 599-8531, Japan
fujita@mecha.osakafu-u.ac.jp

ABSTRACT

The unstable vibration of thin cylindrical shells subjected to axial leakage flows is investigated in this paper. Considering the fluid structure interaction between shells and fluids flowing through a narrow passage, the coupled equation of motion between shells and flowing fluids is derived using the Flügge's shell theory and the Navier-Stokes equation. Focusing on the vibrational characteristics of a thin cylindrical shell with both simply-supported ends, that are the axisymmetric vibration, the lateral vibration and the circumferential vibration, the unstable phenomenon of thin cylindrical shells are clarified by using the root locus based on the complex eigenvalue analysis. The numerical parameter studies on the shells with both simply-supported ends are performed taking the dimensions of shells, the characteristics of flowing fluids and the clearance of an annular gap as parameters. The influences of these parameters on the threshold of the instability of the coupled vibration between a shell and a flowing fluid are discussed.

NOMENCLATURE

a: Radius of the inner thin cylindrical shell.
C: Extensional rigidity.
$C_{a,n}$: Added damping matrix.
D: Flexural rigidity.
H, $\bar{H}$, ΔH: Annulus gap, steady gap, unsteady gap.
h: Thickness of a shell.
$K_{s,n}$, $K_{a,n}$: Structural modal stiffness and added stiffness matrices.
ℓ: Length of a shell.
$M_{s,n}$, $M_{a,n}$: Structural modal mass and added mass matrices.
m: Axial mode number.
m_x, m_θ: Moments acting on a shell.
n: Circumferential mode number.
P, $\bar{P}$, ΔP: Pressure, steady pressure, unsteady pressure.
p_x, p_θ, p_z: Axial, circumferential, radial forces acting on a shell.
Q_x, $\bar{Q}_x$, ΔQ_x: Axial flow rate, steady axial flow rate, unsteady axial flow rate.
Q_y, ΔQ_y: Circumferential flow rate, unsteady circumferential flow rate.
t: Time.
U, V, W: Flow velocity in x-, θ-, z_f-directions ($y = a\theta$).
u, v, w: Displacement of a shell in x-, θ-, z_s-directions ($y = a\theta$).
$\hat{u}_{mn}$, $\hat{v}_{mn}$, $\hat{w}_{mn}$: Eigenmodes of a shell in axial, circumferential and radial directions.
x: Coordinate corresponding to axial direction of a shell.
y: Displacement corresponding to circumferential direction of a shell (y-axis is transformed by $\theta = y/a$).
z_f, z_s: Coordinate corresponding to radial direction of a fluid and a shell.
ζ_{in}, ζ_{out}: Pressure loss coefficients at inlet and outlet.
θ: Angle corresponding to circumferential direction (θ-axis is transformed by $y = a\theta$).
μ: $= \rho_s h$ (mass per unit area).
ν_s: Poisson's ratio.
ν_f: Kinematic viscosity.
$\{\xi_n(t)\}$: Displacement vector of a shell.
$\xi_{mn}(t)$: Modal displacement of a shell.
ρ_s, ρ_f: Density of a shell and a fluid.
ω_{mn}: Eigenfrequency in a vacuum.

INTRODUCTION

A fluid is often conducted as the energy transfer medium in many industrial machinery. On the other hand, many machines and structures must be constructed and operated in flowing fluid. The velocity of fluid becomes faster in order to increase the efficiency and make the size more compact. In this paper, the fluid-elastic instability which occurs in a coaxial cylindrical shell structures subjected to an axial leakage flow is investigated. The machines and structures which are composed of coaxial cylindrical shells are recognized in the combustors of gas turbines, the core internals of nuclear power plants, the annular structures in large diameter pipings and various large scale seals and valves.

The vibrational mechanisms of machines and structures subjected to an axial leakage flow have been studied by many researchers[(1)–(7)]. However, the studies conducted to be based on a shell theory have not reported so many. Accordingly, the further remaining study can be considered to be still necessary for developing the improved coupled vibrational analysis methodology.

In this paper, considering the fluid structure interaction between shells and fluids flowing through a narrow passage, the coupled equation of motion between shells and flowing fluids is derived using the Flügge's shell theory and the Navier-Stokes equation which is applied to an annular leakage flow in accompanying with some approximated assumptions. The vibrational characteristics of a thin cylindrical shell can be classified into the axisymmetric vibration, the lateral vibration and the circumferential vibration. Especially focusing on these three types of vibrations, the unstable phenomenon of a thin cylindrical shell of simply-supported at both ends are clarified by using the root locus due to the complex eigenvalue analysis. The critical fluid velocities for the divergence and the flutter phenomena of a thin cylindrical shell subjected to an axial leakage flow are analyzed and defined. Moreover, comparing with the uncoupled vibration modes, the coupled vibration modes are shown to be considerably changed between before and after the unstable phenomenon. Finally, regarding the dimensions of a thin cylindrical shell, the velocity of an axial leakage flow, the characteristics of flowing fluid and the clearance of an annular gap as the parameters, the physical influences of parameters on the threshold of the instability of the coupled vibration are considered and discussed.

EQUATION OF MOTION

Let us consider the thin cylindrical shells subjected to the axial leakage flow shown in Figure 1. The inner shell is set to be elastic, on the other hand, the outer shell is set to be rigid, and both are set coaxially. On the other hand, even if the inner structure is a rigid cylinder and the outer structure is an elastic shell, this theory can be similarly applied by changing the acting direction of the fluid pressure. The dimensions of the inner elastic shell are length: ℓ, radius: a, thickness: h, the Young's modulus: E, the Poisson ratio: ν_s, density: ρ_s. The leakage flow passes through the annulus between two shells. The inner elastic shell vibrates due to the fluid force of the axial leakage flow. In this paper, the annular gap is considered to be sufficiently small compared with the radius and the length of a shell.

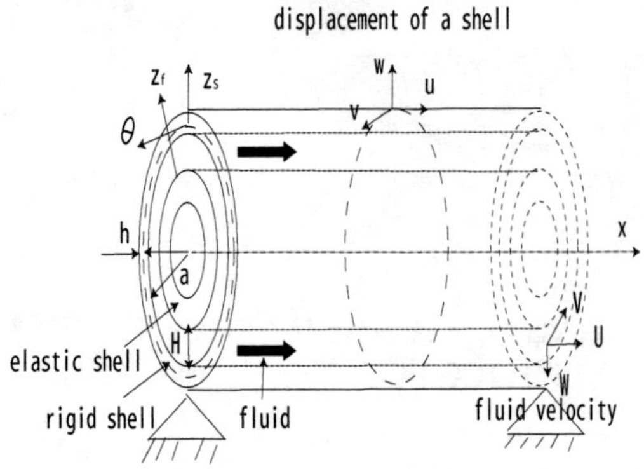

Figure 1. Analytical model of a thin cylindrical shell

The equations of motion of an incompressible fluid based on the Navier-Stokes equations are given by equations of continuity and momentum:

$$\frac{\partial U}{\partial x} + \frac{\partial V}{\partial y} + \frac{\partial W}{\partial z_f} = 0 \tag{1}$$

$$\frac{\partial U}{\partial t} + U\frac{\partial U}{\partial x} + V\frac{\partial U}{\partial y} + W\frac{\partial U}{\partial z_f} = -\frac{1}{\rho_f}\frac{\partial P}{\partial x} + \nu_f\frac{\partial^2 U}{\partial z_f^2} \tag{2}$$

$$\frac{\partial V}{\partial t} + U\frac{\partial V}{\partial x} + V\frac{\partial V}{\partial y} + W\frac{\partial V}{\partial z_f} = -\frac{1}{\rho_f}\frac{\partial P}{\partial y} + \nu_f\frac{\partial^2 V}{\partial z_f^2}, \tag{3}$$

where U, V, W: flow velocity of the x-, θ-, z_f- directions, respectively ($y = a\theta$), P: pressure, ρ_f: density of a fluid and ν_f: kinematic viscosity of a fluid as shown in Figure 1.

The equations of a thin cylindrical shell are given by the Flügge's equations:

$$C\left\{\frac{\partial^2 u}{\partial x^2} + \frac{1-\nu_s}{2a^2}\frac{\partial^2 u}{\partial \theta^2} + \frac{1+\nu_s}{2a}\frac{\partial^2 v}{\partial x \partial \theta} + \frac{\nu_s}{a}\frac{\partial w}{\partial x}\right\}$$
$$+D\left\{\frac{1-\nu_s}{2a^4}\frac{\partial^2 u}{\partial \theta^2} - \frac{1}{a}\frac{\partial^3 w}{\partial x^3} + \frac{1-\nu_s}{2a^3}\frac{\partial^3 w}{\partial x \partial \theta^2}\right\}$$

$$-\mu\frac{\partial^2 u}{\partial t^2}+p_x=0 \tag{4}$$

$$C\left\{\frac{1}{a^2}\frac{\partial^2 v}{\partial\theta^2}+\frac{1-\nu_s}{2}\frac{\partial^2 v}{\partial x^2}+\frac{1+\nu_s}{2a}\frac{\partial^2 u}{\partial x\partial\theta}+\frac{1}{a^2}\frac{\partial w}{\partial\theta}\right\}$$
$$+D\left\{\frac{3(1-\nu_s)}{2a^2}\frac{\partial^2 v}{\partial x^2}-\frac{3-\nu_s}{2a^2}\frac{\partial^3 w}{\partial x^2\partial\theta}\right\}$$
$$-\mu\frac{\partial^2 v}{\partial t^2}+p_\theta-\frac{1}{a}\frac{\partial m_\theta}{\partial\theta}=0 \tag{5}$$

$$-C\frac{1}{a}\left\{\frac{1}{a}\frac{\partial v}{\partial\theta}+\frac{w}{a}+\nu_s\frac{\partial u}{\partial x}\right\}$$
$$+D\left\{\frac{1}{a}\frac{\partial^3 u}{\partial x^3}-\frac{1-\nu_s}{2a^3}\frac{\partial^3 u}{\partial x\partial\theta^2}+\frac{3-\nu_s}{2a^2}\frac{\partial^3 v}{\partial x^2\partial\theta}\right.$$
$$-\frac{\partial^4 w}{\partial x^4}+\frac{2}{a^2}\frac{\partial^4 w}{\partial x^2\partial\theta^2}+\frac{2}{a^4}\frac{\partial^2 w}{\partial\theta^2}$$
$$\left.+\frac{1}{a^4}\frac{\partial^4 w}{\partial\theta^4}+\frac{1}{a^4}w\right\}$$
$$-\mu\frac{\partial^2 w}{\partial t^2}+p_z-\frac{\partial m_x}{\partial x}-\frac{1}{a}\frac{\partial m_\theta}{\partial\theta}=0, \tag{6}$$

where $\mu=\rho_s h$, and p_x, p_θ, p_z: external fluid forces acting on a shell, m_x, m_θ: moments due to the fluid pressure acting on a shell, and the extensional rigidity and the flexural rigidity of a shell C, D are $C=Eh/(1-\nu_s^2)$, $D=Eh^3/12(1-\nu_s^2)$.

FLOW RATE AND BOUNDARY CONDITION FOR FLOW

Now, introducing the axial and the circumferential flow rate $Q_x=\int_0^H U dz_f$, $Q_y=\int_0^H V dz_f$, where H is the gap, and integrating the equations (1)-(3), the flow velocity W disappears by using boundary conditions between a shell and a fluid and the following equations which describe the relation among the flow rate, the pressure and the gap are obtained:

$$\frac{\partial Q_x}{\partial x}+\frac{\partial Q_y}{\partial y}+\frac{\partial H}{\partial t}=0 \tag{7}$$

$$\frac{1}{\rho_f}\frac{\partial P}{\partial x}=-\frac{1}{H}\left\{\frac{\partial Q_x}{\partial t}+\frac{\partial}{\partial x}\left(\frac{Q_x^2}{H}\right)\right.$$
$$\left.+\frac{\partial}{\partial y}\left(\frac{Q_xQ_y}{H}\right)+\frac{12\nu_f Q_x}{H^2}\right\} \tag{8}$$

$$\frac{1}{\rho_f}\frac{\partial P}{\partial y}=-\frac{1}{H}\left\{\frac{\partial Q_y}{\partial t}+\frac{\partial}{\partial x}\left(\frac{Q_xQ_y}{H}\right)\right.$$
$$\left.+\frac{\partial}{\partial y}\left(\frac{Q_y^2}{H}\right)+\frac{12\nu_f Q_y}{H^2}\right\}. \tag{9}$$

The boundary conditions of the pressure at the inlet and outlet are obtained as follows:

$$P(0,y,t)=P_{in}-(1+\zeta_{in})\frac{\rho_f Q_x^2(0,y,t)}{2H^2(0,y,t)} \tag{10}$$

$$P(\ell,y,t)=P_{out}+\zeta_{out}\frac{\rho_f Q_x^2(\ell,y,t)}{2H^2(\ell,y,t)}, \tag{11}$$

where P_{in} and P_{out} are the pressures just in front of the inlet and just behind the outlet in an axial leakage flow, and ζ_{in}, ζ_{out} are the pressure loss coefficients at the inlet and the outlet.

UNSTEADY PRESSURE

The gap, the pressure and the flow rate can be divided into steady and unsteady parts as $H=\bar{H}+\Delta H(x,y,t)$, $P=\bar{P}(x)+\Delta P(x,y,t)$, $Q_x=\bar{Q}_x+\Delta Q_x(x,y,t)$, $Q_y=\Delta Q_y(x,y,t)$, where the bar and Δ imply the steady and unsteady values. The steady circumferential flow rate is neglected because it is considered to be smaller than the axial flow rate. Moreover, $\bar{H}$ and $\bar{Q}_y$ are constant. By applying the perturbation method and neglecting higher order terms of unsteady components, equations (7)-(11) can be linearized as follows:

$$\frac{\partial\Delta Q_x}{\partial x}+\frac{\partial\Delta Q_y}{\partial y}=-\frac{\partial\Delta H}{\partial t} \tag{12}$$

$$\frac{1}{\rho_f}\frac{\partial\Delta P}{\partial x}=-\frac{1}{\bar{H}}\frac{\partial\Delta Q_x}{\partial t}-\frac{\bar{Q}_x}{\bar{H}^2}\frac{\partial\Delta Q_x}{\partial x}-\frac{12\nu_f}{\bar{H}^3}\Delta Q_x$$
$$+\frac{\bar{Q}_x^2}{\bar{H}^3}\frac{\partial\Delta H}{\partial x}+\frac{\bar{Q}_x}{\bar{H}^2}\frac{\partial\Delta H}{\partial t}+\frac{36\nu_f\bar{Q}_x}{\bar{H}^4}\Delta H \tag{13}$$

$$\frac{1}{\rho_f}\frac{\partial\Delta P}{\partial y}=-\frac{1}{\bar{H}}\frac{\partial\Delta Q_y}{\partial t}-\frac{\bar{Q}_x}{\bar{H}^2}\frac{\partial\Delta Q_y}{\partial x}-\frac{12\nu_f}{\bar{H}^3}\Delta Q_y \tag{14}$$

$$\Delta P(0,y,t)=-(1+\zeta_{in})\frac{\rho_f\bar{Q}_x^2}{\bar{H}^2}$$
$$\times\left\{\frac{\Delta Q_x(0,y,t)}{\bar{Q}_x}-\frac{\Delta H(0,y,t)}{\bar{H}}\right\} \tag{15}$$

$$\Delta P(\ell,y,t)=\zeta_{out}\frac{\rho_f\bar{Q}_x^2}{\bar{H}^2}\left\{\frac{\Delta Q_x(\ell,y,t)}{\bar{Q}_x}-\frac{\Delta H(\ell,y,t)}{\bar{H}}\right\}. \tag{16}$$

Eliminating the flow rates from these equations, the relation between the unsteady pressure and the unsteady gap can be derived as follows:

$$\frac{1}{\rho_f}\left(\frac{\partial^2}{\partial x^2}+\frac{\partial^2}{\partial y^2}\right)\Delta P=\frac{1}{\bar{H}}\frac{\partial^2\Delta H}{\partial t^2}+\frac{2\bar{Q}_x}{\bar{H}^2}\frac{\partial^2\Delta H}{\partial x\partial t}$$

$$+\frac{12\nu_f}{\bar{H}^3}\frac{\partial \Delta H}{\partial t}+\frac{\bar{Q}_x^2}{\bar{H}^3}\frac{\partial^2 \Delta H}{\partial x^2}+\frac{36\nu_f \bar{Q}_x}{\bar{H}^4}\frac{\partial \Delta H}{\partial x}. \quad (17)$$

FLUID-SHELL COUPLED VIBRATION DISPLACEMENTS

In this paper, classifying the vibrations into three types of vibrations, which are the axisymmetric, the lateral and the circumferential vibrations, the fluid-structure coupling and the unstable behavior are considered and discussed.

The displacements of a shell subjected to an axial leakage flow are assumed to be approximated by superposing the eigenmode functions in a vacuum. Employing the eigenmodes in the axial, the circumferential and the radial directions, $\hat{u}_{mn}(x)$, $\hat{v}_{mn}(x)$ and $\hat{w}_{mn}(x)$, respectively, the fluid-coupled displacements $u(x,\theta,t)$, $v(x,\theta,t)$ and $w(x,\theta,t)$ are expanded as follows:

$$\begin{pmatrix} u(x,\theta,t) \\ v(x,\theta,t) \\ w(x,\theta,t) \end{pmatrix} = \sum_{m=1}^{\infty}\sum_{n=0}^{\infty} \begin{pmatrix} \hat{u}_{mn}(x)\cos n\theta \\ \hat{v}_{mn}(x)\sin n\theta \\ \hat{w}_{mn}(x)\cos n\theta \end{pmatrix} \xi_{mn}(t), \quad (18)$$

where $\xi_{mn}(t)$ is the function of time. $\hat{u}_{mn}(x)$, $\hat{v}_{mn}(x)$ and $\hat{w}_{mn}(x)$ are obtained by the exact solution.

Solution Satisfying Flügge's Equation

The solution which satisfies equations(4)-(6) is assumed to be expressed in the circumferential direction term, axial direction one and time function separately as follows[8]:

$$\left.\begin{aligned} u &= U'(x)\cos(n\theta)(A\cos\omega t + B\sin\omega t) \\ v &= V'(x)\sin(n\theta)(A\cos\omega t + B\sin\omega t) \\ w &= W'(x)\cos(n\theta)(A\cos\omega t + B\sin\omega t) \end{aligned}\right\}, \quad (19)$$

where ω is eigenfrequency in a vacuum. Substituting equation(19) into equations(4)-(6) and then as U', V', W' in equations(19) are the required mode functions, let consider U', V', W' to be expressed as $U' = c_1 e^{\lambda x}, V' = c_2 e^{\lambda x}, W' = c_3 e^{\lambda x}$. Finaly, the following equation is obtained:

$$e^{\lambda x}\begin{bmatrix} U_1' & U_2' & U_3' \\ V_1' & V_2' & V_3' \\ W_1' & W_2' & W_3' \end{bmatrix}\begin{bmatrix} c_1 \\ c_2 \\ c_3 \end{bmatrix} = \vec{O}, \quad (20)$$

where $U_1' \sim U_3', V_1' \sim V_3', W_1' \sim W_3'$ are functions of ω and λ. The conditions of free vibration in equation(20) is that c_1, c_2, c_3 must have not zero solutions. Therefore,

$$\det\begin{bmatrix} U_1' & U_2' & U_3' \\ V_1' & V_2' & V_3' \\ W_1' & W_2' & W_3' \end{bmatrix} = a_0\lambda^8 + a_1\lambda^6 + a_2\lambda^4 + a_3\lambda^2 + a_4 = 0 \quad (21)$$

has to be satisfied. Here, $a_0 \sim a_4$ are functions of ω. Solving equation(21), $\lambda_j (j = 1\cdots, 8)$ are obtained, then the general solutions concerning U', V', W' are given as follows:

$$\left.\begin{aligned} U'(x) &= \textstyle\sum_{j=1}^{8} c_3{}^{(j)} k_u(\lambda_j)\, e^{\lambda_j x} \\ V'(x) &= \textstyle\sum_{j=1}^{8} c_3{}^{(j)} k_v(\lambda_j)\, e^{\lambda_j x} \\ W'(x) &= \textstyle\sum_{j=1}^{8} c_3{}^{(j)} e^{\lambda_j x} \end{aligned}\right\}, \quad (22)$$

where $k_u(\lambda_j)$ and $k_v(\lambda_j)$ are functions of $c_1/c_3, c_2/c_3$ obtained by solving equation(20).

Boundary Condition of a Shell

When a thin cylindrical shell is assumed to be simply supported at both ends in the radial and circumferential directions, 8 equations of condition can be given. That is, substituting equation(22) into these equations of boundary conditions, the following equation is obtained:

$$Q\left\{c_3{}^{(1)}\cdots c_3{}^{(8)}\right\}^T = 0, \quad (23)$$

where Q is the matrix having 8×8 elements which are functions of ω and λ. The solutions of λ in equation(21) are complex solutions consisting of complex conjugates. Therefore, the following vector $\vec{d}$ consisting of 16 elements is introduced.

$$\vec{d} = [Re(c_3^{(1)})\ldots Re(c_3^{(8)})\ Im(c_3^{(1)})\ldots Im(c_3^{(8)})]^T. \quad (24)$$

Equation(23) is modified by using $\vec{d}$ as follows:

$$\left[Re(Q) \vdots -Im(Q)\right]\vec{d} = \vec{0}. \quad (25)$$

Equation(25) is expressed by a matrix of 8×16. Besides, $[c_3^{(1)}\ldots c_3^{(8)}]^T$ in equation(23) are complex conjugates because λ in equation(23) have complex conjugates. Then, as the real parts are equal and the imaginary parts have equal absolute values but not equal, the 8 following equations are given:

$$[R]\vec{d} = \vec{0} \quad (26)$$

Combining equations(25) and (26) together, 16 equations are obtained as follows:

$$[S]\vec{d} = \vec{O}, \tag{27}$$

where

$$S = \begin{bmatrix} Re(Q) & \vdots & -Im(Q) \\ & R & \end{bmatrix}. \tag{28}$$

For the condition that $\vec{d}$ have nontrivial solutions, the following equation must be satisfied:

$$\det[S] = 0. \tag{29}$$

Solving the simultaneous equations consisting of equations (21) and (29), ω and λ can be obtained. Furthermore, substituting ω and λ into eqnuation (27), vector $\vec{d}$ can be given by getting the nontrivial solutions. Following the above-mentioned procedure, ω_{mn} *and* λ_{mn} in the m-th axial mode and the n-th circumferential mode can be obtained. When $A_{mnj} = d_j + \mathbf{i}d_{j+8}$($\mathbf{i}$ is $\sqrt{-1}$) using vectors $\vec{d}$ determined by ω_{mn} and λ_{mn} is set, the eigenmode functions in the m-th axial mode and the n-th circumferential mode are expressed finally as follows:

$$\left.\begin{aligned} \hat{u}_{mn}(x) &= \textstyle\sum_{j=1}^{8} A_{mnj}k_u(\lambda_{mnj})\, e^{\lambda_{mnj}x} \\ \hat{v}_{mn}(x) &= \textstyle\sum_{j=1}^{8} A_{mnj}k_v(\lambda_{mnj})\, e^{\lambda_{mnj}x} \\ \hat{w}_{mn}(x) &= \textstyle\sum_{j=1}^{8} A_{mnj}e^{\lambda_{mnj}x} \end{aligned}\right\}. \tag{30}$$

Besides, for the cylindrical shell simply supported at both ends, the analytical eigenmode functions in a vacuum can be considered to be assumed as $\hat{u}_{mn}(x) = C_{1mn}\cos(m\pi x/\ell)$, $\hat{v}_{mn}(x) = C_{2mn}\sin(m\pi x/\ell)$, $\hat{w}_{mn}(x) = C_{3mn}\sin(m\pi x/\ell)$ which satisfy the simply-supported boundary condition. In fact, when the above-assumed eigenmode function is adopted in numerical simulation, the almost same results are obtained.

COUPLED FLUID FORCE

Here, in this section, the unsteady pressure ΔP is derived and fluid forces and moments are calculated more specifically. Since the unsteady gap width ΔH coincides with the radial displacement of a shell, the following relation can be obtained: $\Delta H = w$. Using this relation and equation(18), equation(17) is rewritten as follows:

$$\frac{1}{\rho_f}\left(\frac{\partial^2}{\partial x^2} + \frac{\partial^2}{\partial y^2}\right)\Delta P(x,\theta,t)$$

$$= \sum_{m=1}^{\infty}\sum_{n=0}^{\infty}\cos n\theta\Big\{A_{mn}(x)\ddot{\xi}_{mn}(t)$$

$$+B_{mn}(x)\dot{\xi}_{mn}(t) + C_{mn}(x)\xi_{mn}(t)\Big\}, \tag{31}$$

where the coefficients $A_{mn}(x)$, $B_{mn}(x)$, $C_{mn}(x)$ are the functions given by the modal function $\hat{w}_{mn}(x)$ and its derivatives. Since y is described as $y{=}a\theta$, the left hand side of equation(31) is the function of x, θ and t. The unsteady pressure ΔP is assumed to be written as $\Delta P(x,\theta,t) = \sum_{n=0}^{\infty}\Delta P_n(x,t)\cos n\theta$, where ΔP_n means the unsteady pressure at $\theta = 0$. Substituting this into equation(31), the ordinary differential equation is obtained. Solving the obtained ordinary differential equation under the linearlized boundary conditions based on equations(10) and (11), the relation between the unsteady pressure $\Delta P(x,\theta,t)$ and the modal displacement is obtained:

$$\Delta P(x,\theta,t) = \sum_{m=1}^{\infty}\sum_{n=0}^{\infty}\cos n\theta\Big\{\bar{m}_{mn}(x)\ddot{\xi}_{mn}(t)$$

$$+\bar{c}_{mn}(x)\dot{\xi}_{mn}(t) + \bar{k}_{mn}(x)\xi_{mn}(t)\Big\}. \tag{32}$$

The functions $\bar{m}_{mn}(x)$, $\bar{c}_{mn}(x)$ and $\bar{k}_{mn}(x)$ are coefficient functions. Their concrete forms are omitted.

FLUID-SHELL COUPLED EQUATION AND STABILITY ANALYSIS

The fluid forces acting on a shell are given by $p_x = -\mu_f\partial\Delta U/\partial z_f|_{z_f=0}$, $m_x = \mu_f\partial\Delta U/\partial z_f|_{z_f=0}(h/2)$, $p_\theta = -\mu_f\partial\Delta V/\partial z_f|_{z_f=0}$, $m_\theta = \mu_f\partial\Delta V/\partial z_f|_{z_f=0}(h/2)$, $p_z = \Delta P(x,\theta,t)|_{z_f=0} = \Delta P(x,\theta,t)$. Assuming the velocity distribution as $\Delta U = (1/2\mu_f)\partial\Delta P/\partial x z_f(z_f - \bar{H})$, $\Delta V = (1/2\mu_f)\partial\Delta P/\partial\theta z_f(z_f - \bar{H})$, the fluid forces become $p_x = -(\bar{H}/2)\partial\Delta P/\partial x$, $p_\theta = (\bar{H}/2)\partial\Delta P/\partial\theta$, $p_z = \Delta P(x,\theta,t)$, $m_x = (h\bar{H}/4)\partial\Delta P/\partial x$, $m_\theta = (h\bar{H}/4)\partial\Delta P/\partial\theta$, where $\mu_f = \rho_f\nu_f$.

Substituting these and the equation(18) into the equations of a thin cylindrical shell (4)-(6) and utilizing the orthogonality of the mode of the axial and the circumferential directions, the following fluid-shell coupled equation of motion is obtained, that is, multiplying equations(4)-(6) by the eigenfunction of the p-th in the axial direction and the q-th in the circumferential direction and integrating the resultant equation along the axial direction (0 to ℓ) and the circumferential direction (0 to 2π):

$$\ddot{\xi}_{pq}(t) + \omega_{pq}^2\xi_{pq}(t) = -\sum_{m=1}^{\infty}\Big\{\hat{m}_{pmq}\ddot{\xi}_{mq}(t)$$

$$+\hat{c}_{pmq}\dot{\xi}_{mq}(t) + \hat{k}_{pmq}\xi_{mq}(t)\Big\}, \quad (33)$$

where ω_{pq} is the eigenfrequency of the shell in a vacuum. The left hand side of the above equation is described on only the shell, therefore it can be understood that the left hand side is orthogonalized with the p-th axial mode and the q-th circumferential mode. On the other hand, the fluid forces in the right hand side in equation(33) can not be orthogonalized. That is, the combination of the function of time with $m = 1, 2, \cdots, \infty$ is left. By this equation, even in the fluid-shell coupled vibration subjected to the axial leakage flow, the vibration modes in the axial direction and the circumferential one can be transformed to the product of the function of each variable when a coupled eigenvalue analysis is performed for equation(33) and the orthogonality condition among the coupled modes can be applicable. Therefore, it is seen that similarly to the modes in a vacuum, the axisymmetric, the lateral and the circumferential vibrations are considered separately even in a fluid-shell coupling. Truncating the number of mode as M, employing vectors and matrices form, and changing q to n, equation(33) can be rewritten as follows:

$$[M_{s,n} + M_{a,n}]\{\ddot{\xi}_n(t)\} + [C_{a,n}]\{\dot{\xi}_n(t)\} + [K_{s,n} + K_{a,n}]\{\xi_n(t)\} = \{0\} \quad (34)$$

where
$$\begin{aligned}
\{\xi_n(t)\} &= \{\xi_{1n}(t), \xi_{2n}(t), \cdots, \xi_{Mn}(t)\}^T \\
[M_{s,n}] &= I_{M\times M} \\
[K_{s,n}] &= \mathrm{diag}\{\omega_{1n}^2, \omega_{2n}^2, \cdots, \omega_{Mn}^2\} \\
[M_{a,n}] &= [\hat{m}_{mpn}]_{1\le m,p\le M} \\
[C_{a,n}] &= [\hat{c}_{mpn}]_{1\le m,p\le M} \\
[K_{a,n}] &= [\hat{k}_{mpn}]_{1\le m,p\le M}
\end{aligned}$$

The matrices $[M_{s,n}]$, $[K_{s,n}]$ are the modal mass and stiffness of only a shell, and $[M_{a,n}]$, $[C_{a,n}]$, $[K_{a,n}]$ are the added mass, added damping and added stiffness matrices of a fluid in a gap, and n=0, n=1 and $n \ge$2 denote the axisymmetric, the lateral and the circumferential vibrations respectively. By calculating the complex eigenvalue s and plotting the root locus, the stability of a system can be evaluated.

Here, there are $2M$ complex eigenmodes of equation(34) for fixed n. M pairs of eigenvalues exist. Each pair is a set of complex conjugate numbers or a pair of two real numbers. Let us call the pairs the $1, 2, 3, \cdots M$-th eigenvalues. When the imaginary part of a complex eigenvalue is not zero, the complex eigenvalue with a positive imaginary part in a pair of complex conjugate eigenvalues is adopted. On the other hand, when the imaginary part is zero, the eigenvalue with a larger real part in a pair of eigenvalues is adopted. Thus, the focused M complex eigenvalues are written as $\{\lambda_{1,n}, \lambda_{2,n}, \cdots, \lambda_{M,n}\}$. Let the complex eigenvectors corresponding to $\lambda_{k,n}$ be $\{\Psi_{k,n}\}$, and the m-th element of $\{\Psi_{k,n}\}$ be $\Psi_{k,n,m}$. Moreover, let $\eta_{k,n}(t)$ be the coupled function of time corresponding to $\{\Psi_{k,n}\}$. Therefore, the coupled k-th complex eigenmodes are expressed as follows:

$$\left.\begin{aligned}
u_{k,n}^c(x,\theta) &= \sum_{m=1}^{M} \hat{u}_{mn}(x)\cos(n\theta)\Psi_{k,n,m} \\
v_{k,n}^c(x,\theta) &= \sum_{m=1}^{M} \hat{v}_{mn}(x)\sin(n\theta)\Psi_{k,n,m} \\
w_{k,n}^c(x,\theta) &= \sum_{m=1}^{M} \hat{w}_{mn}(x)\cos(n\theta)\Psi_{k,n,m}
\end{aligned}\right\} \quad (35)$$

Here, note the coupled complex eigenvalues and eigenmodes vary when the flow velocity varies. By employing these equations, the displacement of a cylindrical shell can be expressed as follows:

$$\begin{pmatrix} u(x,\theta,t) \\ v(x,\theta,t) \\ w(x,\theta,t) \end{pmatrix} = \sum_{k=1}^{M}\sum_{n=0}^{N} \begin{pmatrix} u_{k,n}^c(x,\theta) \\ v_{k,n}^c(x,\theta) \\ w_{k,n}^c(x,\theta) \end{pmatrix} \eta_{k,n}(t) \quad (36)$$

The fact that the imaginary part of complex eigenvalue s satisfies $Im(s) \neq 0$ and the real part $Re(s) > 0$ implies dynamically unstable (flutter) and $Im(s) = 0$ and $Re(s) > 0$ implies statically unstable (divergence). According to these procedures, the critical flow velocity and complex eigenmode shape for the leakage-flow induced vibration can be evaluated.

NUMERICAL ANALYSIS

Dimensions for Numerical Analysis Studies

As numerical analysis examples, let us consider the inner shell is elastic, and the outer shell is rigid like a hollow rigid cylinder as shown in Figure 1. Two types of shell models are considered here: One is the shorter elastic shell, the other is the longer one comparing with the radius and the length shown in Table 1. The elastic shell is assumed to be simply supported at both ends in the radial and circumferential directions as explaind above. The fluid is water, and the elastic shell is considered to be made of steel. Hereafter, let us call these models as Case 1 and Case 2.

Table 1. Dimensions of numerical models

	Case 1	Case 2
length [m]	0.5	2
radius [m]	0.5	0.5
thickness [m]	0.002	0.002
gap width [m]	0.002	0.002
ℓ/a	1	4
a/h	250	250

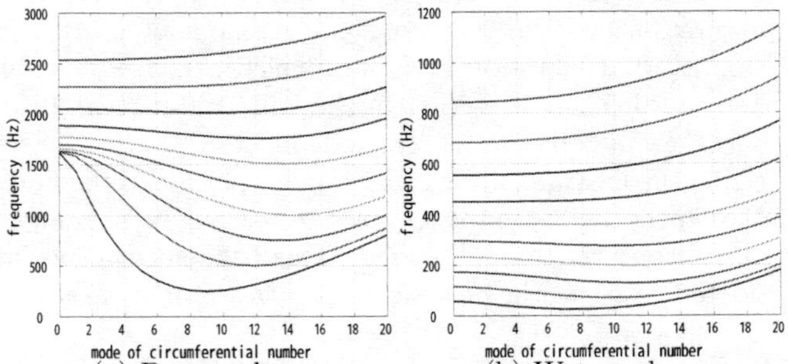

(a) Dry mode (b) Wet mode

Figure 2. Eigenvalue analysis in Case 1

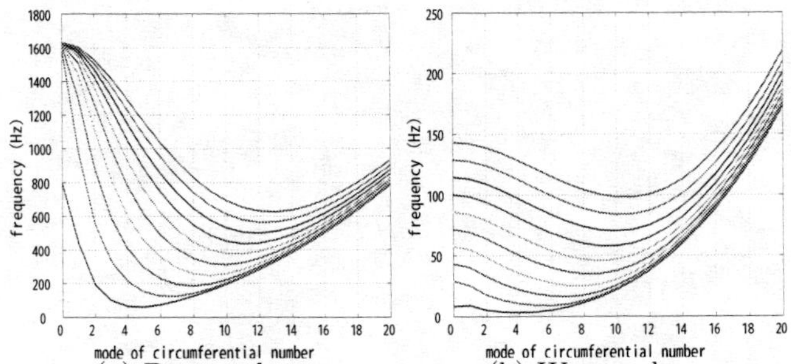

(a) Dry mode (b) Wet mode

Figure 3. Eigenvalue analysis in Case 2

Eigenvalue Analysis of Numerical Models

Figures 2 and 3 show the eigenfrequencies in dry and wet conditions for Case 1 and Case 2 respectively. Their curves show the eigenfrequencies of the m-th axial mode from lower number $m = 1, 2, \cdots 10$ in order. The dry condition means the eigenvalues in a vacuum, and the wet means those in a still fluid. Hereafter let us call them dry modes, and wet modes respectively. When a fluid is air, wet modes almost equal the dry ones except for a small annulus gap. In these figures, the axial mode number m is taken as a parameter. As a fluid is water, the eigenfrequencies of wet modes become much lower due to the added mass effect of the fluid. In Case 1, the eigenfrequencies of dry modes show the minimum at the mode $(m, n) = (1, 9)$. On the other hand, those of wet modes show that at the mode $(m, n) = (1, 8)$. In Case 2, the eigenfrequencies of dry modes show the minimum at the mode $(m, n) = (1, 5)$. On the other hand, those of wet modes show that at the mode $(m, n) = (1, 4)$.

Stability Analysis

Concerning the axisymmetric, the lateral and the circumferential vibrations of a thin cylindrical shell subjected to an axial leakage flow, their dynamic stability is investigated in this subsection.

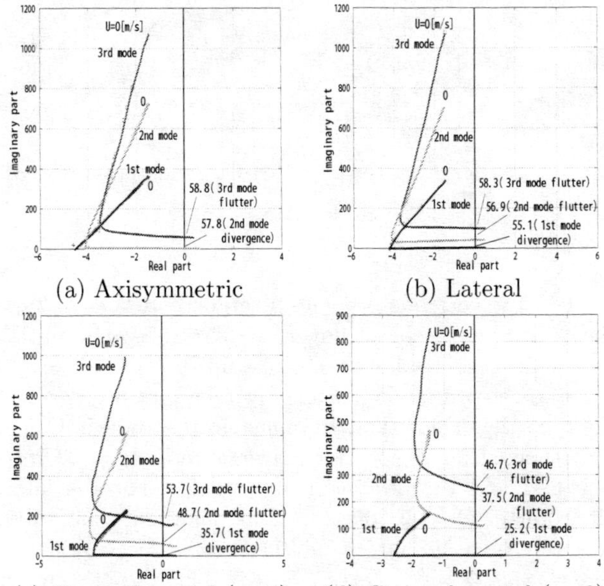

(a) Axisymmetric (b) Lateral
(c) Circumferential (n=4) (d) Circumferential (n=8)

Figure 4. Root locus in Case 1

Figures 4 and 5 show the root loci of three types of vibrations for Case 1 and Case 2 respectively. Regarding the axisymmetric vibration in Case 1, the 1st axial mode is stable, the 2nd mode goes into divergence at $U = 57.8m/s$ and the 3rd mode goes into flutter at $U = 58.8m/s$, then the 3rd mode goes into divergence at $U = 60.0m/s$ increasing the fluid velocity. For the lateral vibration in Case 1, the 1st axial mode becomes divergence at $U = 55.1m/s$, the 2nd mode becomes flutter at $U = 56.9$m/s and the 3rd mode becomes flutter at $U = 58.3m/s$, then the 2nd and the 3rd mode become divergence at $U = 57.3m/s$, $U = 59.6m/s$ respectively. Moreover, for the 4th circumferential vibration in Case 1, the 1st axial mode becomes divergence at $U = 35.7$m/s, the 2nd mode becomes flutter at $U = 48.7m/s$ and the 3rd mode becomes flutter at $U = 53.7m/s$, and then the 2nd and the 3rd mode become divergence. For the 8th circumferential vibration in Case

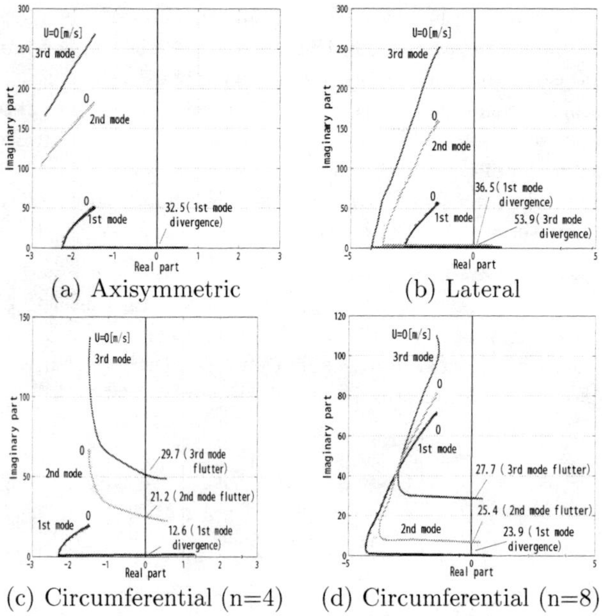

(a) Axisymmetric (b) Lateral

(c) Circumferential (n=4) (d) Circumferential (n=8)

Figure 5. Root locus in Case 2

1, the 1st axial mode becomes divergence at $U = 25.2m/s$, the 2nd mode becomes flutter at $U = 37.5m/s$ and the 3rd mode becomes flutter at $U = 46.7m/s$, and then the 2nd and the 3rd mode become divergence respectively. On the other hand, for the axisymmetric vibration in Case 2, the 1st axial mode goes into divergence at $U = 32.5m/s$. The 2nd and the 3rd mode are stable. For the lateral vibration in Case 2, the 1st axial mode goes into divergence at $U = 36.5m/s$ and the 3rd mode goes into divergence at $U = 53.9m/s$. Further, for the 4th circumferential vibration in Case 2, the 1st axial mode becomes divergence at $U = 12.6m/s$, the 2nd mode becomes flutter at $U = 21.2m/s$ and the 3rd mode becomes flutter at $U = 29.7m/s$ and then they become divergence respectively. For the 8th circumferential vibration in Case 2, the 1st axial mode becomes divergence at $U = 23.9m/s$, the 2nd mode becomes flutter at $U = 25.4m/s$ and the 3rd mode becomes flutter at $U = 27.7m/s$ and then they become divergence respectively. In both cases, the circumferential vibration initially goes into the unstable phenomenon. Concerning with the minimum critical velocity, the numerical analysis must be performed for many modes of the circumferential vibration. However, the determination that which circumferential vibration mode initially goes into the unstable phenomenon is found to be due to the dimensions of a shell.

Complex Eigenmode

In this subsection, comparing with the uncoupled vibration modes, the coupled vibration modes are shown to have changed considerably between before and after the unstable phenomenon. Figures 6, 7 and 8 show the mode shapes at every instance, $t = 0, T/7, \cdots, 6T/7, T$ over one period T. Figure 6 shows the 2nd complex eigenmode of the lateral vibration in Case 1 which becomes unstable. Figure 6 (a) indicates that in the still fluid at U=0m/s, and Figure 6 (b) the flutter mode at $U = 56.9m/s$. The complex eigenmode at U=0m/s is almost the same as the dry mode. However, the flutter mode shape at 56.9m/s is fairly different from at $U = 0m/s$. Figure 7 shows the 1st complex eigenmode of the axisymmetric vibration in Case 2. Figure 7 (a) indicates that in the still fluid at $U = 0m/s$, and Figure 7 (b) the divergence mode at $U = 32.5m/s$. The right hand side of Figure 7 (b) is the one side-view of the divergence mode. Figure 8 shows the 2nd complex eigenmode of the 4th circumferential vibration in Case 2 which becomes unstable. Figure 8 (a) indicates that in the still fluid at $U = 0m/s$, and Figure 8 (b) the flutter mode at $U = 21.2m/s$. Especially, the location of the maximum amplitude of the flutter mode over one period is seen not to be fixed axially and to move along the axial direction. This behavior can be considered to mean the generation of a traveling wave axially.

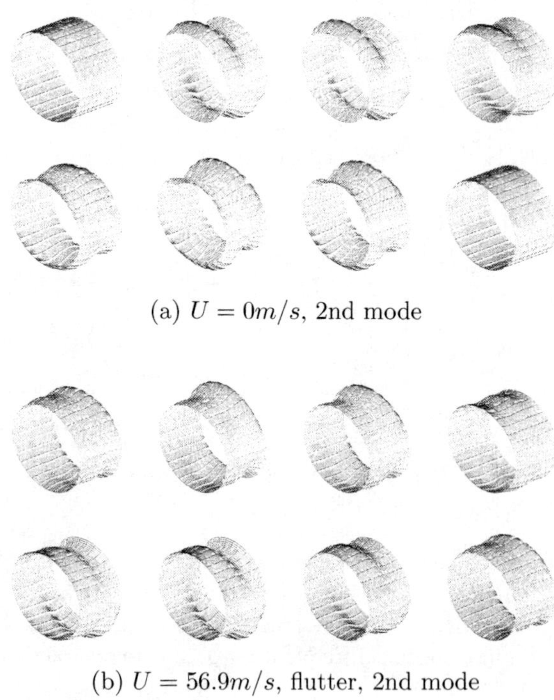
(a) $U = 0m/s$, 2nd mode

(b) $U = 56.9m/s$, flutter, 2nd mode

Figure 6. Complex eigenmode of the lateral vibration (Case 1, 2nd mode, n=1)

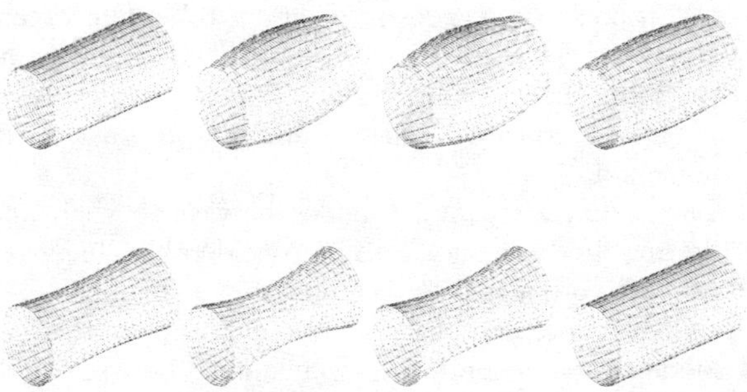

(a) $U = 0m/s$, 1st mode

(b) U=32.5m/s, divergence, 1st mode

Figure 7. Complex eigenmode of the axisymmetric vibration (Case2, 1st mode, n=0)

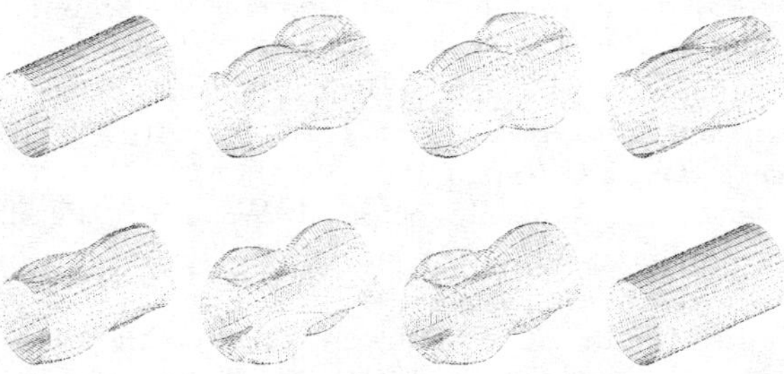

(a) $U = 0m/s$, 2nd mode

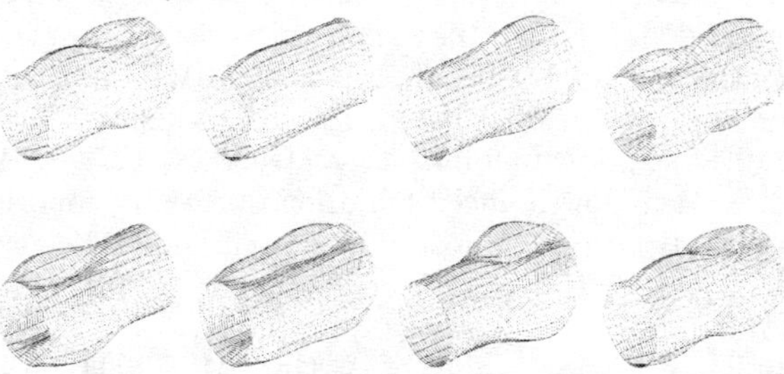

(b) $U = 21.2m/s$, flutter, 2nd mode

Figure 8. Complex eigenmode of the circumferential vibration (Case 2, 2nd mode, n=4)

Influence of Gap Width

In this following, the influence of the gap width on the stability is examined. Figure 9 (a), (b) and (c) show the influence of the gap width on the critical velocity of the divergence phenomenon of the axisymmetric vibration, the flutter of the lateral vibration and the flutter penomenon of the 8th circumferential vibration in Case 1 respectively. Figure 9 (d), (e) and (f) show the influence of the gap width on the critical velocity of the divergence phenomenon of the axisymmetric vibration, the divergence phenomenon of the lateral vibration and the flutter phenomenon of the 4th circumferential vibration in Case 2 respectively. The small gap width can be seen to make the vibration of a thin cylindrical shell subjected to an axial leakage flow more unstable.

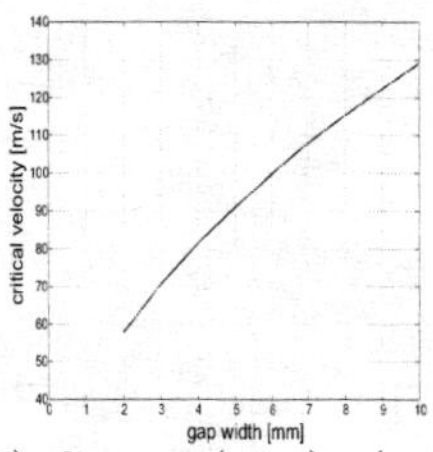

(a) Case 1,(m,n)=(2,0) divergence

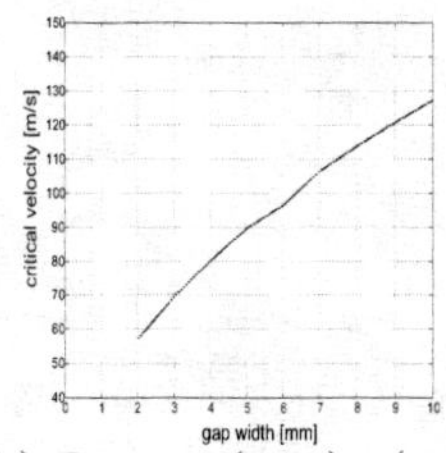

(b) Case 1, (m,n)=(2,1) flutter

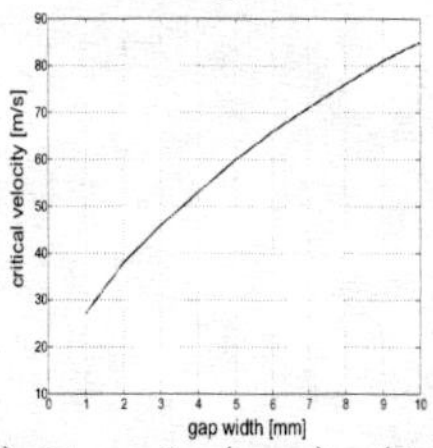

(c) Case 1 ,(m,n)=(2,8) flutter

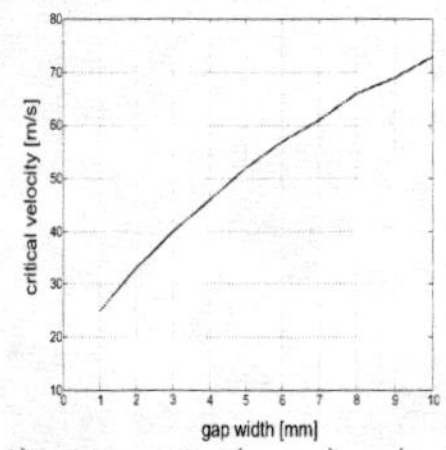

(d) Case 2, (m,n)=(1,0) divergence

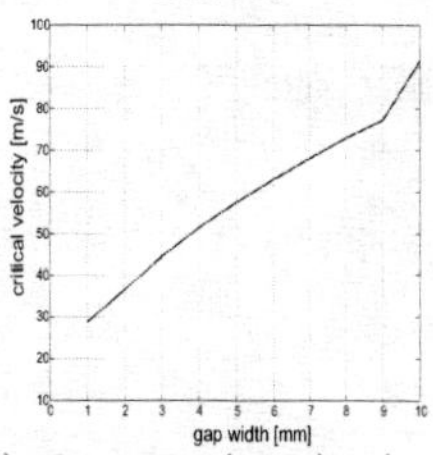

(e) Case 2, (m,n)=(1,1) divergence

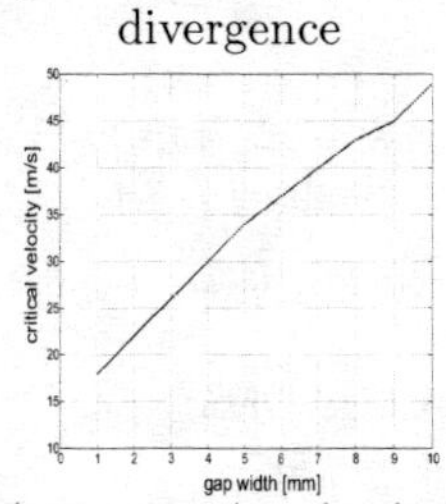

(f) Case 2, (m,n)=(2,4) flutter

Figure 9. Influence of the gap width on the critical velocity

Influence of Kinematic Viscosity

In this following, the influence of the kinematic viscosity of a fluid is examined. Figure 10 shows the root loci of the 4th circumferential vibration in Case 2. Figure 10 (a) and (b) indicate the root locus of $\nu_f = 1.004 \times 10^{-6} m^2/s$

corresponding to water at the temperature $T' = 20°C$, $\nu_f = 0.4747 \times 10^{-6} m^2/s$ at $T' = 60°C$ and $\nu_f = 0.2945 \times 10^{-6} m^2/s$ at $T' = 100°C$ of the 1st and the 2nd axial mode respectively. The real part which means the damping term in the equation of motion shifts to the unstable right side when the temperature becomes high. Figure 11 (a) and (b) show the influence of the kinematic viscosity on the critical velocity of the 1st and the 2nd mode respectively. On the contrary, the critical velocities of the 1st and the 2nd axial modes become high slightly when the temperature becomes high.

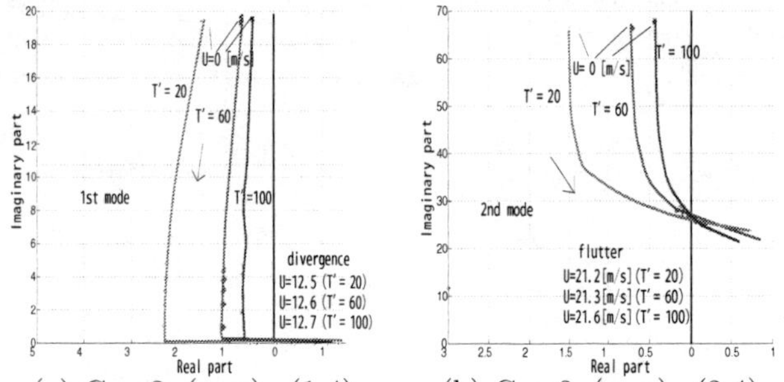

(a) Case2, (m,n)=(1,4) divergence

(b) Case2, (m,n)=(2,4) flutter

Figure 10. Influence of the kinematic viscosity on the root locus of the circumferential vibration

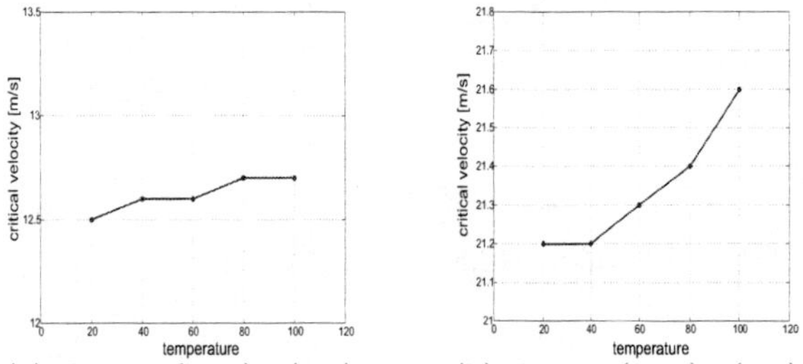

(a) Case2, (m,n)=(1,4) divergence

(b) Case2, (m,n)=(2,4) flutter

Figure 11. Influence of the kinematic viscosity on the critical velocity

CONCLUSIONS

The following conclusions are obtained:

(1) The coupling vibrational analysis method of a thin cylindrical shell subjected to an axial leakage flow is proposed by using the Flügge's shell theory and the Navier-Stokes equation.

(2) Focusing on the three types of vibration, that are the axisymmetric, the lateral and the circumferential vibration of a thin cylindrical simply-supported shell subjected to an axial leakage flow, the unstable phenomena are clarified by using the root locus based on the complex eigenvalue analysis.

(3) From numerical parameter studies, the circumferential vibration initially goes into the unstable phenomenon. However, the dimensions of a shell are found to have the influence on the determination that which circumferential vibration mode goes first into the unstable phenomenon.

(4) The coupled vibration modes between a shell and a flowing fluid are shown to be considerably changed between before and after the unstable phenomena. Especially the location of maximum amplitude of the flutter mode over one period is seen not to be fixed and to move along the axial direction. A traveling wave can be found to generate axially.

(5) The smaller gap width is found to make a thin cylindrical shell subjected to an axial leakage flow more unstable.

(6) When the temperature of water becomes high, that means the kinematic viscosity and the density of water is changed, the real part in root locus shifts to the unstable right side. On the contrary, the critical velocity becomes high slightly.

REFERENCES

(1) Hobson, D. E. (1982), Fluid-elastic instabilities caused by flow in an annulus, *Proceedings of BNES 3rd Int. Conf. in Vibration in Nuclear Plant*, 440-461.

(2) Païdoussis, M. P. (1966), Dynamics of flexible slender cylinders in axial flow, part 1. theory, part 2. experiment, *Journal of Fluid Mechanics, 26*, 717-751.

(3) Chen, S. S., & Wambsganss, M. W. (1972), Parallel-Flow-Induced Vibration of Fuel Rods, *Nuclear Eng. and Design, 18*, 253-278.

(4) Fujita, K., Ito, T. & Baba, K. (1993), Study on the unstable vibration of cylindrical shells due to leakage flow, *Proc. of the Int. Conf. on Aero-Hydroelasticity*, 286-295.

(5) Fujita, K., & Shintani, A., (2001), Axial leakage flow-induced vibration of the elastic rod as the axisymmetric continuous flexible beam, Trans. of ASME, *Journal of Pressure Vessel Technology, 123-4*, 421-428.

(6) Fujita, K., & Shintani, A., & Ono, M, (2003), Axial leakage flow-induced vibration of a thin cylindrical shell with respect to axisymmetric vibration, Trans. of ASME, J. of Pressure Vessel Thechnology, 125-2, 151-157.

(7) Fujita, K., & Shintani, A,, & Ono, M., (2003), Axial leakage flow-induced vibration of a thin cylindrical shell with respect to lateral vibration, Trans. of ASME, J. of Pressure Vessel Thechnology, 125-2, 158-164.

(8) Fujita, K., & Saito, A., (2002), Coupled Vibration Analysis of a Cylindrical Shell with Liquid, Piping and Unaxisymmetrical Structures, Proc. of ASME, ESDA2002/APM-026, CD-ROM, 1-10.

Proceedings of IMECE04
2004 ASME International Mechanical Engineering Congress and Exposition
November 13-20, 2004, Anaheim, California USA

IMECE2004-60634

Control Valve Design Impact on Piping Vibration

James McGhee
Principal Consultant
JMDynamics Limited
New Bridgegate House
69 Dixon Road
Glasgow G42 8AT
Scotland, UK
Ph 44-(0)141-424-1447
Fax 44-(0)141-424-1447
Email – jmcghee@JMDynamics.com

Doug Newlands
Controls Engineer
Email – newlanD@bp.com
Stuart Farquhar
Mechanical Engineer
farquhS@bp.com
Ph 44-1224-832000
Fax 44-1224-832550
BP Exploration
Fairburn Industrial Estate
Dyce, Aberdeen AB21 7PB
Scotland, UK

Herbert L. Miller
Valve Consultant
Control Components Inc.
22591 Avenida Empresa
Rancho Santa Margarita,
CA 92688
Ph 949-858-1877
Fax 949-858-1878
hlm@ccivalve.com

ABSTRACT

Vibration of the recycle piping system on the Main Oil Line (MOL) Export Pumps from a platform in the North Sea raised concern about pipe breakage due to fatigue. Failures had already occurred in associated small bore piping and the instrument air supply lines and control accessories on the recycle flow control valves. Concern also existed due to the vibration of non-flowing pipe work and systems such as the deck structure, cable trays and other instrumentation, which included fire and gas detection systems. Many changes involving bracing of small bore attachments, stiffening of supports, adding supports and stiffing the deck structure were implemented without resolving the problem. The vibration was finally solved by adding enough pressure stages to assure the valve trim exit velocities and energy levels were reduced to levels demonstrated historically as needed in severe service applications. This vibration energy reduction was more than 16 times. This was achieved by reducing the valve trim exit velocity from peaks of 74 m/s to 12 m/s (240 ft/s to 40 ft/s).

Key words: Control Valves, Piping, Vibration, Design

INTRODUCTION

In the traditional design process there are five steps taken to assure a control valve will meet the conditions specified. These are:

1. Size for the flow capacity required.
2. Select the body that will house the trim, provide pressure containment and satisfy the inlet and outlet connections.
3. Select the materials for the valve trim and body compatible with the fluid conditions.
4. Account for the fluid stream. (That is; is the design acceptable for the applicable cavitation, noise, material erosion and/or flashing fluid conditions that may exist?)
5. Ensure the valve is properly actuated.

This is where many control valve manufacturers stop, however there is a sixth step, Miller [1], needed to assure the valve would meet the application requirements and provide good control with high reliability and minimum maintenance. This sixth step is:

6. Verify that the fluid exiting from the valve trim orifices does not produce jets that will cause undue forces.

These turbulent forces will affect the control and reliability of the valve and the system in which it is installed.

Control of the trim exiting jets is accomplished by providing enough staging during the pressure letdown so that kinetic energy levels meet an established criterion. The energy criterion has been presented in a number of references [1 – 4] and is further discussed in this paper as it applies to the Main Oil Line (MOL) recycle control valve application, first reported by Chapman et. al. [5].

This MOL application further illustrates an example in which reducing the trim exit kinetic energy levels inside the valve significantly impacted the piping vibration. The application is the recycle valves for the MOL pumps on a platform in the North Sea. Providing a valve in which the kinetic energy was maintained below the criterion resolved a potentially serious vibration problem where fatigue of the pressurized piping could have caused failure.

The Main Oil Line Recycle System:

There are two parallel pumps used to export oil from the platform. Each pump has a recycle line to ensure minimum flow conditions for the pump are met when the overall flow rates are below the normal operating levels and to protect the pumps during start-up and shut-down. The parallel recycle lines are shown in Figure 1. The recycle lines return the oil from the discharge side of the pumps to a cooler on the suction side and then the oil flows back through the pumps. Due to the severe duty on these valves (high flow rates and high pressure differential) high levels of vibration in the recycle valves and their associated pipework had been observed during startup and shut down and under different flow bias conditions existing between the two pumps.

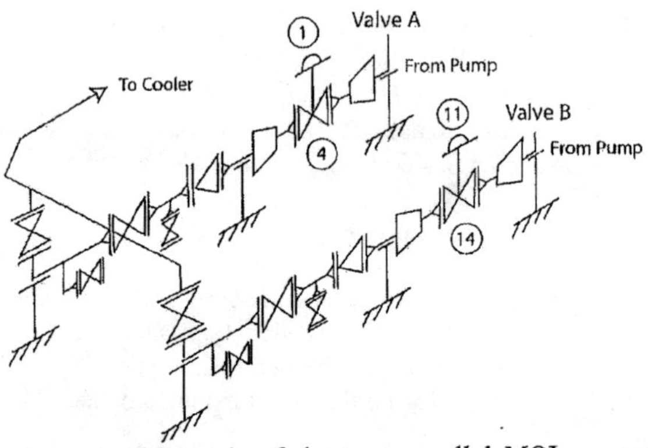

Figure 1. Schematic of the two, parallel MOL pump recycle systems showing points of valve vibration measurement.

These high vibration levels had resulted in repeated fatigue failures during commissioning and early life operation of the installation. The failures included weld failures at small bore connections, failures of the electronic position controllers on the valves and failures of various instrument tubing runs associated with the pumps and the discharge pipework. The levels of vibration in the module steelwork in the area surrounding the pumps also caused concern due to transmission to other non-associated equipment and the plant.

In order to mitigate the effects of the vibration in the short term, various remedial actions involving bracing of all small bore attachments on the recycle lines, stiffening of supports, additional main line supports and stiffening of the local deck structure were implemented. These modifications were partially successful in reducing the vibration levels of the pipework in the area of the recycle valves. However, the source of the vibration remained and continued failures of small instrumentation piping illustrated that the original policy of trying to contain the effects of the vibration would not provide the acceptable level of operation required by the platform operator in the long term.

Possible Solutions:

There are three accepted methods of addressing problems of this nature. The first two methods address the symptoms of the problem and try to contain the vibration that results from the excitation source. They are to increase the system damping and to increase the system stiffness. The third method attempts to minimize the excitation at the source and is, therefore, the preferred method.

Regarding the damping approach there are a number of very effective dampers available for pipework systems that will absorb the vibrational energy from the system and turn it into heat. These dampers act at all frequencies up to approximately 60 Hz and are extremely effective at the low frequency end of the spectrum.

One problem with this approach is that the source of vibration is still present and the potential for valve problems still exists. Another problem is that while these dampers are very effective in controlling the pipework vibration response, significant vibrational energy is still imparted into the structure through the reactions at the damper connecting points and if a high degree of structural stiffness is not achieved at the base of the dampers, significant reductions in damping efficiency occur. This was a particular issue for the MOL pump recycle valves as there was no suitable structural steel in the vicinity of the recycle pipework for a rigid mounting of the dampers. A major construction project would have been needed to provide a structure of sufficient stiffness to allow the dampers to be effective.

Increasing the piping system stiffness in situations of broad band excitation and high amplitude/low frequency response is beneficial since the natural frequencies of the system are moved to higher values. At higher frequencies the broadband excitation levels are lower, and it is more difficult for the excitation to produce vibration levels that would result in failure. As in the case of the damper approach, however, the level of vibration in the valve would be unchanged and the potential for valve problems such as control and instrumentation failure and gland packing assembly failure would still exist.

In the case of the MOL recycle valves the location and orientation of the pipework, made increasing the stiffness extremely difficult and only minor improvements would have been possible. These minor changes would not have significantly altered the pipework vibration response and would not therefore have been beneficial.

The valve trim should be selected to keep the kinetic energy below these levels. Levels of energy are expressed using the same units as pressure and the equation for energy is also the expression for dynamic pressure. The kinetic energy in the above equation is a relative energy in that it is the energy per unit volume.

Table 1. Valve Trim Outlet Kinetic Energy Criteria				
Service Conditions	**Kinetic Energy Criteria**	Velocity if Oil at 0.777 Sp. Gr.	Velocity if Water*	Velocity if 7 MPa (1000 psig) air*
	kPa (psi)	m/s (ft/s)	m/s (ft/s)	m/s (ft/s)
Continuous Service, Single Phase Fluids	**480 (70)**	35 (115)	30 (100)	105 (345)
Cavitating and Multi-phase Fluid Outlet	**275 (40)**	27 (87)	23 (75)	----
Vibration Sensitive System	**75 (11)**	14 (46)	12 (40)	42 (140)

* Noise, cavitation, and erosion requirements may dictate lower velocities.

The preferred method is to address the source and reduce the levels of excitation. If the overall level of energy being produced within the valve is not decreased there will always be the potential for vibration problems to exist in the system. Previous solutions to this type of problem have included:

> Changing the flow parameters, (e.g. reduce flow velocities by increasing pipework diameter) or altering system pressures to reduce the pressure drop. Or, alter the energy coupling mechanisms within the valve by changing the trim characteristics to low vibration type. Pressure across these valve designs is gradually reduced by staging to reduce the level of energy exiting the trim.

Energy and Velocity Levels

The fluid kinetic energy exiting a control valve trim has a significant impact on the behavior of the downstream piping system. Experience has shown that control of this parameter can significantly reduce excessive vibration in existing piping systems. The kinetic energy is defined as follows.

$$KE = \frac{\rho V^2}{2M}$$

where, M = 1000 for metric or 4636.8 for imperial units

and KE is in kPa or psi, V is in m/s or ft/s, ρ is in kg/m^3 or lb/ft^3

The velocity in this expression is the average trim outlet velocity and 'ρ' is the density of the fluid at the exit. Different valve trims and the location within the trim for the velocity and other considerations in the calculation of the kinetic energy are expanded upon Borden [2]; ISA's Control Valve guide.

Velocity criteria for liquids are much lower than for gases because liquid densities are much higher, resulting in higher energy levels. While the velocity limits are quite different, the kinetic energy limits are the same.

The valve trim exit kinetic energy criteria, Table 1, for control valve applications are also shown in the ISA guide, Borden [2].

For the recycle valve application the conservative design criterion is to use the 480 kPa (70 psi) level. This application is usually governed by cavitation and a final check needs to be made to be sure the energy is less than values in the table to minimize a lot of bubble formation. Cavitation damage is usually not an issue with oil. With crude oil there is a likelihood of degassing of some of the more volatile components and then liquefying of the bubble, a phenomenon quite like cavitation. The duty cycle for recycle valves generally is considered as continuous service because of their frequent use and especially because of their criticality to pump protection. However, in the case of the MOL recycle valves vibration was known to be a problem with the system so that energy level much closer to the vibration sensitive systems criterion of 75 kPa (11 psi) was needed. The need for a quiet system was very important.

Table 2. Valve Sizing Process Conditions	
Flow Rate	404000 kg/h (890000 lb/h)
Inlet Pressure	22 MPa (3200 psi)
Outlet Pressure	2 MPa (280 psi)
Pressure Drop	20 MPa (2920 psi)
Temperature	63 C (145 F)
Specific Gravity	0.777
Required Capacity	32.3K_v (37.8 C_v)
Provided Capacity	35.5 K_v (41.5 C_v)

The Recycle Valve Design

The original recycle valve design absorbed the pressure drop from 22to 2 MPa (3200 to 280 psi) with a normal flow rate of 404000 kg/h (890000 lb/h) using a single cage with large flow paths as the trim. The design conditions are shown in Table 2. This valve was later replaced with a new body and six stage, concentric drilled hole cage trim of overlapping and partially offset holes. The multiple cages were in contact so there was essentially no fluid pressure recovery and no axial flow between stages. The valve was a linear travel globe style body configuration. Unacceptable vibration levels continued even with the six stage trim valve design.

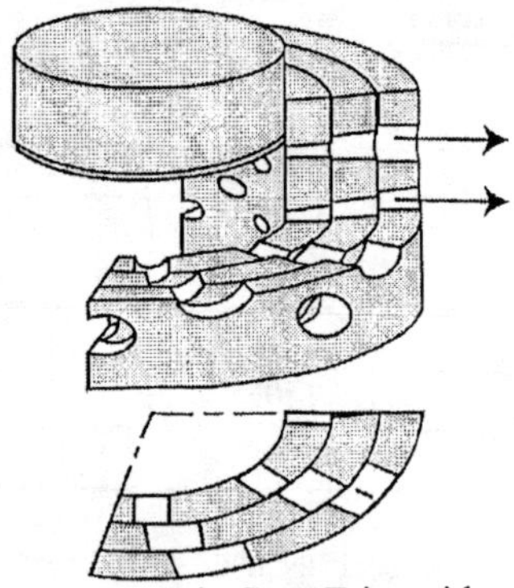

Figure 2. Concentric Cage Trim with Over Lapping holes.

It is also noted that although there are six concentric cages in this trim, it in effect, is less than six stages of pressure drop. This is because the expansion of flow area between stages exceeds the ability of the fluid to follow the turns. A typical concentric cage type trim (three stages) is shown in Figure 2. In effect there is significant jetting through the trim more closely following the line of sight. This is equivalent to a bypassing of the stages and results in less pressure drop than would be associated with a distinct right angle turn.

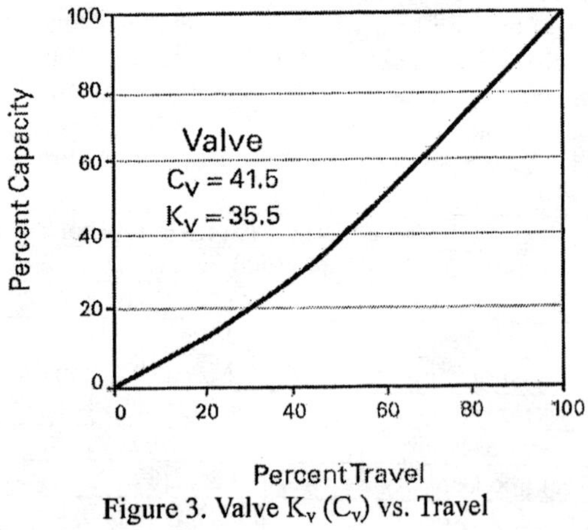

Figure 3. Valve K_v (C_v) vs. Travel

The final valve design was a 22 and 26 stage tortuous path trim consisting of right angle turns to achieve a controlled pressure letdown of the fluid. The resulting valve K_v (C_v) versus travel is shown in Figure 3. The 26 stage trim is in the lower travel and extends for the first 44 percent of the travel and 32 percent of the capacity. This combination of staging resulted from packaging as much resistance in the trim as possible to meet the application flow conditions while using the same valve body. The packaging of additional resistance is maximized after all other design objectives have been satisfied. The design objective was to first be less than the continuous service energy criterion of Table 1 and as close to the vibration sensitive criterion as possible. (The vibration sensitive criterion is very conservative and is usually reserved for control valves used in flow test loop applications where excess turbulence results in added signal noise to flow and/or pressure measurements.) With this trim the outlet velocity and thus the kinetic energy of the jets leaving the trim is maintained within guideline levels. A typical schematic of this type of trim is shown in Figure 4.

The body is a 2500 # ASME globe style with 8 inch hubs to connect with the 8 inch piping. This is the same body type, class, and size that were used for the 6 concentric cage trim. The flow direction is "over the plug" (flow to close) in both cases. The same size valve body allowed easy substitution of the new valve design.

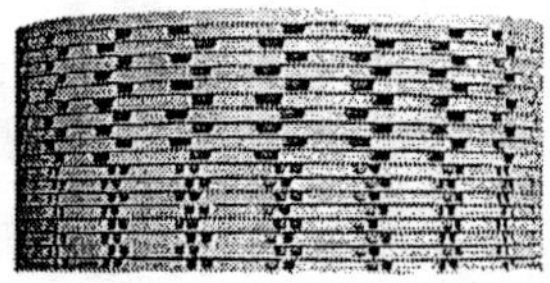

Figure 4. Typical Characterized Multi-Stage Multi-Path Valve Trim

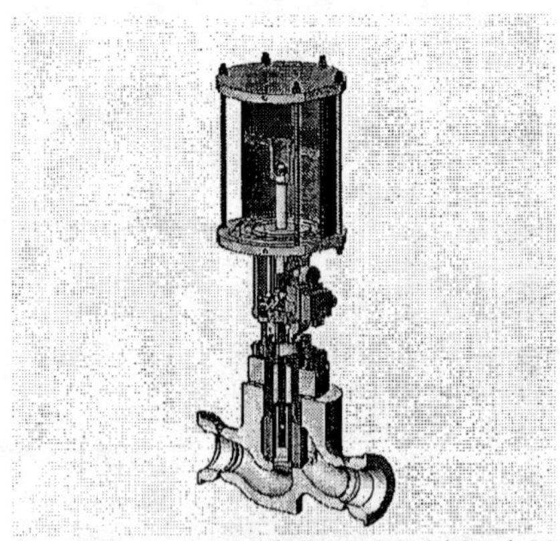

Figure 5. Multi-Stage Multi-Path Valve Assembly Cross-section.

From a vibration standpoint the new multi-stage multi-path trim valve only differs from the previous valve installation in the number and design of the letdown staging. The two bodies are about the same weight; 800 kg (1750 lbs) versus 605 kg (1330 lbs) for the new valve versus the previous valve. Thus the piping system vibration characteristics are essentially the same.

The new multistage valve is taller since it uses a piston style actuator with a spring to open instead of a diaphragm type actuator. This larger height would amplify the piping vibration a bit more in comparison to the shorter diaphragm actuator. A scaled model of the final valve design is shown in Figure 5. The spring for fail open is not shown in Figure 5.

Table 3. Trim Exit Oil Values for the Three Valves

Valve Trial	Valve Trim	Kinetic Energy MPa (psi)	Exit Velocity m/s (ft/s)
1	Single Stage Orifice	20.33 (2950)	227 (745)
2	6 Concentric Cages	2.14 (310)	74 (240)
3	26 Right Angle Turns 22 Right Angle Turns	0.059 (8.5) 0.095 (13.8)	12.2 (40) 15.5 (51)
-	Outlet Pipe 8 inch 34mm thick (1.34 inch)	0.026 (3.8)	8.2 (27)
-	Vibration Sensitive Criterion	0.075 (11)	14 (46)
-	Continuous Criterion	0.480 (70)	35 (116)

A comparison of the three valve designs from a trim energy and velocity standpoint is shown in Table 3 for the design conditions shown in Table 2. Table 3 also presents the downstream piping as well as the Vibration Sensitive and Continuous Service Criteria from Table 1 for comparison.

As shown in Table 3 the change to a 26 stage right angle turn trim resulted in a reduction of energy of over 35 times while the velocity of the exiting jets was reduced by six times when compared to the six concentric cylinder trim. The energy levels resulting from the 26 stage trim also are below the criterion for Vibration Sensitive Systems.

Table 4: Comparison of A and B valve actuators and bodies rms vibration velocity measurements.

Description	Location		July 1999 – 6 Stage Trim	Oct. 2001 22-26 Stage	
			Valve A 100% Open B-0%	Test 1 A-100%, B-60%	Test 2 B-100%, A-10%
Valve A Actuator	1	E/W	24.7	2.0	2.1
		N/S	17.4	3.1	4.0
		Vertical	-	1.2	1.6
Valve A Body	4	E/W	8.6	0.7	0.8
		N/S	10.3	1.4	1.4
		Vertical	18.6	1.3	1.0
Valve B Actuator	11	E/W	12.5	2.9	2.6
		N/S	16.1	4.4	3.8
		Vertical	-	1.2	1.0
Valve B Body	14	E/W	6.8	0.9	0.8
		N/S	8.0	0.9	1.0
		Vertical	-	0.8	0.7

Vibration Measurements and Results

The layout of the recycle pipework and the vibration measurement locations on the valves are shown in Figure 1. Vibration measurements were recorded on the top of the actuator and both recycle valves bodies (locations 1, 4, 11, 14). Other vibration measurements were made throughout the pipeworks; however these four points experienced the highest readings and are most indicative of the severity.

The most severe vibration condition for the valves had previously been identified as the start-up regime with the recycle valves fully open. The principle aim was therefore to assess the vibration levels in this condition for both valves in turn.

With both MOL pumps operational, two principle tests were performed. The first test involved setting Valve A at the 100% open while the parallel Valve B settled under automatic control to the 60% open position. The Pump A discharge pressure was 17.8 MPa (2580 psia) and Pump B discharge was 21.5 MPa (3120 psia). The second test set Valve B 100% open while the parallel Valve A settled automatically at the 10% position. The Pump A discharge pressure was 21.8 MPa (3160 psia) and Pump B discharge was 20.8 MPa (3015 psia). These conditions are very representative of the valve design flow conditions.

The changes in valve position during the two tests occurred too quickly to allow a set of measurements to be recorded during the changes of position, however not too quickly to observe any phenomena that may have been associated with any other valve positions within the range of movement observed.

The rms vibration velocity levels recorded are shown in Table 4 for the July 1999 6-stage trim tests and the October 2001 22-26 stage trim for comparison. As can be seen, the vibration levels recorded for the right angle turn tortuous trim are very low. A maximum level of 4.4 mm/s rms (0.17 in/s) was recorded on the Valve B actuator (north-south) during Test 1 and a maximum of 4.0 mm/s rms (0.16 in/s) recorded on the Valve A actuator (north-south) during Test 2. These measurements constitute a reduction in vibration of a factor of 10 from the previous valves as measured in the east/west direction at the actuator and in the east/west and north/south direction on the body of Valve A. This is a remarkable reduction in vibration level when comparing the valves under the same duty but is even more impressive since the load on the new valve was actually significantly higher than the July 1999 load (higher throughput and larger pressure drop across valve), and yet this order of magnitude reduction was still achieved.

The vibration levels for the B valve from the July 1999 tests are the result of transmission through the deck from the parallel A pipework only. There was no flow through the B pipework. That is, the vibration of Valve B during the 1999 test was caused by flow through Valve A and the subsequent transmission through the deck to Valve B. Comparing these previous results with the recent test, it is seen that reductions in vibration for Valve B range from a factor of 4 to a factor of 9. These results are equally remarkable in that the vibration levels experienced by the valve are lower now with full recycle flow than they were with the previous valve being excited by transmission only.

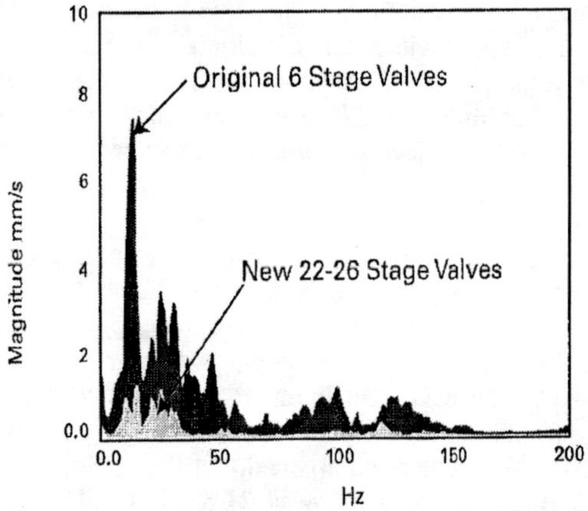

Figure 6. Valve-A Actuator Spectra; A=100%, B=60%

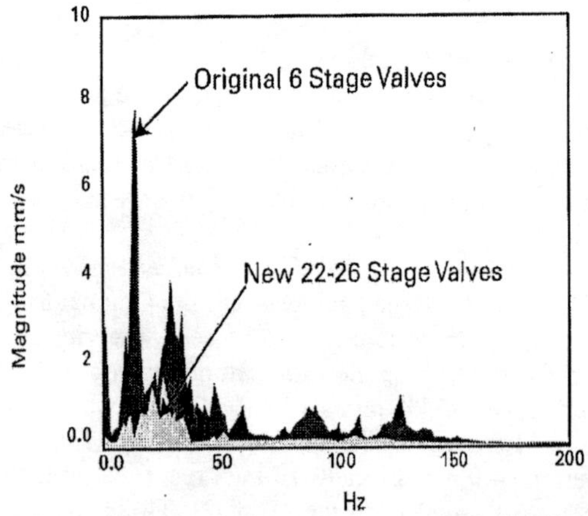

Figure 7. Valve-B Actuator Spectra; B=100%, A=10%

The vibration spectra for the A valve actuator measurements (north/south) for Test 1 is shown in Figure 6 with a comparison of the July results. A similar comparison of the spectra for the B valve actuator (North/South) is shown in Figure 7. All of the spectra show a significant reduction in vibration across the whole frequency range and in particular, the previous strong low frequency response of the actuator has been all but eliminated.

The major spike in the spectra at 13 Hz for the 1999 results is likely the natural frequency of the piping system with other peaks, at 26 and 31 Hz representing the natural frequency of the actuator relative to the valve body and the plug/piston/spring portion of the valve assembly.

These actuator frequency values are quite close to test observations and calculations on a number of similar valves. These three peaks are also visible, but not as distinctive, in the 2001 testing and with significantly reduced magnitudes that represent the more quiet system with the new multi-stage multi-path 22 and 26 turn trim. Certainly the low frequency 13 Hz peak component was the main concern because the pipework and support structures transmitted the resulting vibration throughout the area with damaging consequences.

From Table 3 it is seen that the excessive piping vibration problem was resolved with a trim exit kinetic energy level only about 3.5 times that of the kinetic energy level in the downstream piping. We would have expected there to be no vibration problems using the continuous energy criterion for exit energy level of Table 3. The continuous criterion results in about 20 times the energy in the pipe. The much more conservative criterion of 0.075 MPa (11 psi) was the guideline used in this application because a vibration problem already existed.

Conclusion

The results of the vibration measurement discussed in this paper present another example of the impact of reducing the control valve trim exit energy level on piping systems. As with the other examples noted in the references the maintenance of the energy level below specific criterion will have this affect. Almost all of the piping vibration problems related to control valves that are encountered in the field can be avoided by adding this check of valve trim exit energy level as part of the control valve sizing and design steps. In addition, many other flow related piping vibration problems can be avoided by minimizing the degree of vibration created within the valve and the level of energy leaving it. It is such an easy and mundane check by the valve manufacturer that to not do this puts the user at risk of expensive problem resolution as part of the start up activities of any facility. For those cases in which an existing facility has control valve induced vibration a retrofit of valve trim or replacement of the entire valve assembly, as in this case, can usually be accomplished.

REFERENCES

1. Miller, H. L., 2002, "Trim Exit Kinetic Energy in Control Valves," 2002 Valve World Conference and Exhibition – Flow Control for the Future, Maastricht, November 12-14, KCI Publishing B. V., Zutphen, The Netherlands, pp. 74-79.

2. Borden, G., Jr., 1998, Control Valves – Practical Guides for Measurement and Control, International Society for Measurement and Control, Research Triangle Park, N. C., Chapter 12 – Control Valve Applications, pg. 411-477.
3. Miller, H. L., and Stratton, L. R., 1997, "Fluid Kinetic Energy as a Selection Criteria for Control Valves", ASME Fluids Engineering Division, Summer Meeting, Paper FEDSM97-3464, Vancouver, British Columbia, Canada, June 22-26.
4. Miller, H. l., "Vibration causes and correction in control valves," Valve World, Volume 6 Issue 5, KCI Publishing B. V., Zutphen, Netherlands, pg. 30-35.
5. Chapman, G., McGhee, J., Nash, A., and Selzer, O., 2002, "Heavy vibration MOL recycle valves replaced with severe-service design," Offshore Engineer, Atlantic Communications LLC, Houston, May issue.

Proceedings of IMECE04
2004 ASME International Mechanical Engineering Congress and Exposition
November 13-20, 2004, Anaheim, California USA

IMECE2004-61914

CFD STUDIES ON BURNER SECONDARY AIR FLOW

Anil K. Purimetla
Graduate Student, Mechanical Engineering Department
Jie Cui
Assistant Professor, Mechanical Engineering Department
Stephen Idem
Professor, Mechanical Engineering Department
Sastry Munukutla
Professor, Mechanical Engineering Department, Director, Center For Electric Power
Tennessee Technological University
Cookeville, Tennessee 38505, U.S.A.

ABSTRACT

In many fossil power plants operating today, there is insufficient means to assure the proper balancing of the secondary airflows between the individual burners of wall-fired units in addition there is a problem of dust deposition on the floor. This mismatch leads to decreased boiler efficiency and increased emissions. In this study, a Computational Fluid Dynamics (CFD) modeling of a fossil power plant wind box scale model is performed using the commercial software CFX5.6. The model solves the three dimensional Reynolds averaged Navier-Stokes equations with the K-epsilon turbulence model. The CFD results are validated by the experimental data taken from a $1/8^{th}$ scale model of a wall fired fossil unit. Simulations under various flow conditions are obtained to identify the optimum design in terms of the equalization of the secondary airflow through the burners.

1. INTRODUCTION

An uneven distribution of coal, primary and secondary airflow between the burners of a furnace results in nonuniform combustion of the coal [1]. One means of providing more uniform combustion is to make an accurate balancing of the secondary airflow to the individual burners. There is no standard method commonly used by the power industries for secondary airflow measurements. Two distinct techniques of secondary airflow balancing, both of which can be categorized as cold air velocity testing, are currently used. The cold air refers to the testing and balancing during unit shutdown with only ambient temperature airflow taking place. In the first technique, the measurements are made inside the wind box near the entrance to each individual burner. This method is inaccurate and exposes individuals to high-speed air currents that may bring about unsafe and uncomfortable work conditions. The presence of the person conducting the teat disturbs the flow pattern. In the second technique, an articulated probe is used to perform cold air testing. Thus is quite time consuming. Both techniques are used when the whole plant is not in operation.

In order to take a better approach to secondary air balancing, scale model tests on a $1/8^{th}$ scale model of the wind box and burner arrangement have been conducted. The scale model was subjected to a series of tests to obtain flow measurements. In the mean time, numerical simulation of the secondary airflow balancing within the scale model using CFD has been developed and validated.

The main objective of the current study is to develop a reliable numerical tool that can be used in the balancing of the secondary airflow distribution to allow the burners to perform as designed and to achieve the lowest possible emissions. Various flow conditions are considered in the CFD simulations to bring a near equal flow through the burners.

NOMENCLATURE

ρ	Density, Kg/m^3
μ	Dynamic Viscosity, $N\text{-}sec/m^2$

μ_{eff}	Effective Viscosity, N-sec/m^2
μ_t	Eddy Viscosity, N-sec/m^2
ε	Dissipation rate, m^2/sec
σ_k	Constant (=1.0)
σ_ε	Constant (=1.3)
$C_{\varepsilon 1}$	Constant (=1.44)
$C_{\varepsilon 2}$	Constant (=1.92)
C_μ	Constant (= 0.09)
K	Turbulent kinetic energy, m^2/sec^2
P	Pressure, Pa
U_i	Cartesian velocity components, m/sec
x_i	Coordinate axes
MFR	Mass Flow Rate, Kg/sec

2. CFD MODELING

2.1. GEOMETRY AND MESHING

The present geometry involves the wind box, burner, baffle and porous plates. There are two principal difficulties associated with the numerical modeling of the geometry. The first is that the geometry of the model ranges from large open volumes of the wind box to many small openings of the burners through which the entire flow must pass. The second is the rather large size of the entire model. In the scale model tests, there were 18 burners inside the wind box. While in the numerical simulations, only 9 burners were considered in view of the flow and geometrical symmetry about the central vertical plane. This has greatly reduced the computational load without losing any physical significance in a steady RANS type simulation. Figure 1 shows the typical model of the wind box with nine burners, a baffle and two porous plates.

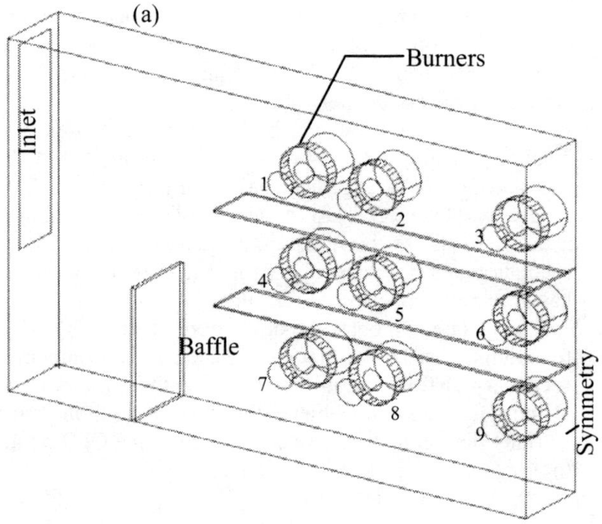

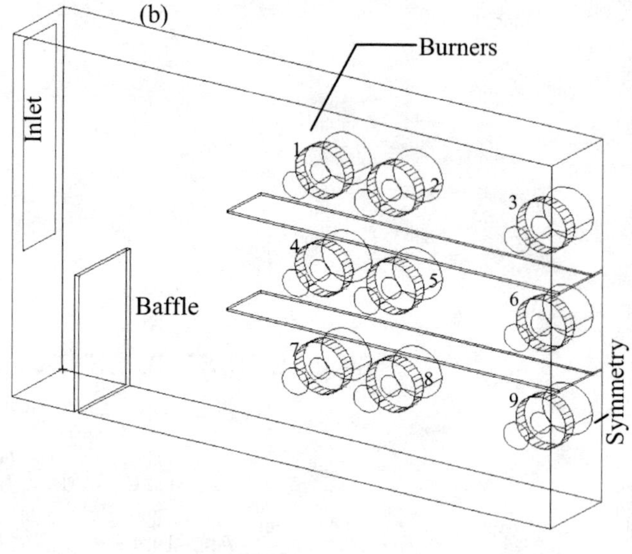

Figure 1. Isometric view of the basic geometry considered for numerical solution (a) baffle position 1; (b) baffle position 2.

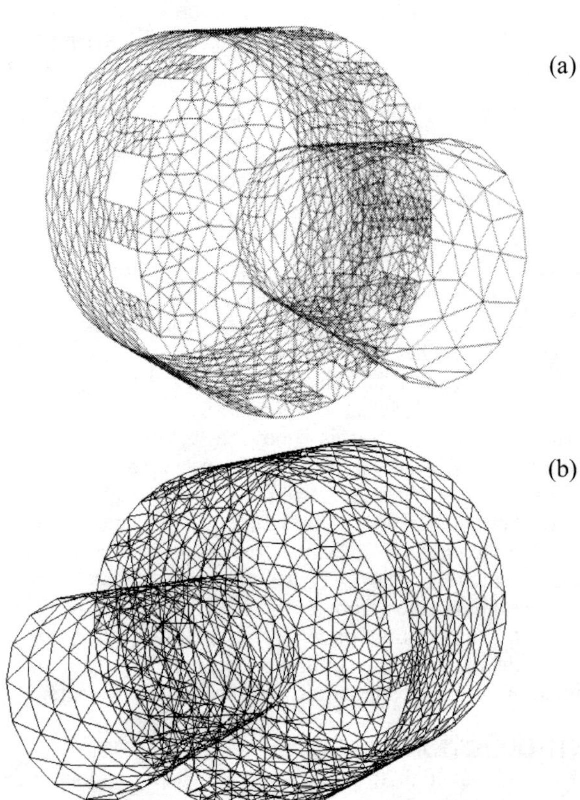

Figure 2. Three-dimensional view of the burner with surface mesh (a) full opening; (b) half opening.

Figure 2 shows the solid model representation of a typical scale model burner. A carefully detailed modeling of the burner was created in order to obtain the desired accurate results. The truncated cone at the front of the burner is a representation of the piping system that caries pulverized coal suspended in primary air to the combustion chamber. In the present calculation model the primary air and coal were not considered. The fourteen rectangular openings on the larger cylinder represent the ports through which the secondary air enters the burner. In order to achieve a balancing of the secondary airflow, a sleeve damper is used to adjust its flow area. For the present study two cases (full and half opening of the rectangular ports) were considered. Each case was calculated at two different positions of the baffle. The baffle positions are shown in Figure 1, the second baffle position is in the midway between the inlet plane and the first baffle position.

The numerical model must contain enough volumes to accurately simulate the flow patterns, especially near the nine burners. The generation of surface mesh and volume mesh is managed by various controls. The control depends on the type of mesh that is used.

2.2. ASSUMPTIONS IN THE CFD APPROACH

The following assumptions were considered for the numerical solution of the wind box with associated geometry using CFD.

- The normal secondary air temperature during the operation is at approximately 600 ^{0}F. However, in the scale model tests, ambient air was used as the fluid. In order to make a legitimate comparison between the simulation and measurements, air at 25°C was considered in the computations. Density variation is negligible and not included in the present study.
- The thickness of the rectangular ports and sleeve damper were not considered because of the geometry complexities.
- In this model instead of porous plates, two plates with equal area are used in the numerical model due to meshing issues.
- Steady-state simulation was assumed for all numerical simulations.

2.3. FLUID MODELS

Since the scale model tests were performed under different conditions, the fluid properties were slightly different. For the air port full opening case, air at 25°C is considered with a viscosity of 1.831×10^{-5} Kg/m s, density of 1.185 kg/m^3 and for half open case the values specified for air are $\mu=1.817\times10^{-5}$ Kg/m s, $\rho = 1.618$ kg/m^3. For all the flow conditions the flow in the wind box is turbulent. In the present numerical analysis K-ε Model is used to solve the turbulence models. In this model both velocity and length scales are solved using separate transport equations. K is the turbulence kinetic energy and is defined as the variance of the fluctuations in velocity. ε is the turbulence eddy dissipation that is the rate at which the velocity fluctuations dissipate.

2.4. GOVERNING EQUATIONS

CFD or numerical modeling is the process of approximating the governing flow conservation equations (e.g., continuity, momentum and energy) for solution on computers. Partial differential equations are approximated by algebraic equations applicable over a small region called a volume or element. The mass and momentum transport equations for steady and incompressible flow in CFX5.6 are [2]

$$\frac{\partial(\rho U_i)}{\partial x_i} = 0 \tag{1}$$

$$\frac{\partial(\rho U_i U_j)}{\partial x_j} = -\frac{\partial P}{\partial x_i} + \frac{\partial}{\partial x_j}(\mu_{eff}\frac{\partial U_i}{\partial x_j} + \mu_{eff}\frac{\partial U_j}{\partial x_j}) \tag{2}$$

where repeated indices imply summation from 1 to 3, ρ is the density, U_i are the Cartesian velocity components, P is the pressure, x_i are the coordinate axes, and μ_{eff} is the effective viscosity, it is defined as

$$\mu_{eff} = \mu + \mu_t \tag{3}$$

where

$$\mu_t = C_\mu \rho \frac{K^2}{\varepsilon} \tag{4}$$

C_μ is a constant and is equal to 0.09, μ_t is the eddy viscosity, K is the turbulent kinetic energy, μ is the dynamic viscosity, and ε is the dissipation rate. The turbulence quantities are calculated using K-ε model.

$$\frac{\partial \rho K}{\partial t} + \frac{\partial}{\partial x_j}(\rho K U_j) - \frac{\partial}{\partial x_j}(\frac{\mu_{eff}}{\sigma_k}\frac{\partial K}{\partial x_j})$$

$$= \mu_t \frac{\partial U_i}{\partial x_j}(\frac{\partial U_i}{\partial x_j} + \frac{\partial U_j}{\partial x_i}) - \frac{2}{3}\frac{\partial U_j}{\partial x_j}(\mu_t \frac{\partial Ui}{\partial xj} + \rho K) - \rho\varepsilon \tag{5}$$

$$\frac{\partial \rho\varepsilon}{\partial t} + \frac{\partial}{\partial x_j}(\rho\varepsilon U_j) - \frac{\partial}{\partial x_j}(\frac{\mu_{eff}}{\sigma_\varepsilon}\frac{\partial \varepsilon}{\partial x_j})$$

$$= \frac{\varepsilon}{K}(C_{\varepsilon 1}(\mu_t \frac{\partial U_i}{\partial x_j}(\frac{\partial U_i}{\partial x_j} + \frac{\partial U_j}{\partial x_i}) - \frac{2}{3}\frac{\partial U_j}{\partial x_j}(\mu_t \frac{\partial U_i}{\partial x_j} + \rho K)) - C_{\varepsilon 2}\rho\varepsilon) \tag{6}$$

Where σ_k=1.0, σ_ε=1.3, $C_{\varepsilon 1}$=1.44, $C_{\varepsilon 2}$=1.92 are constants from K-ε model.

2.5. BOUNDARY CONDITIONS

The following boundary conditions were applied to the model:

- Inlet Boundary Condition with a specified Mass Flow Rate (MFR) = 0.178 Kg/s, in two other cases, half and twice this mass flow rate were also defined.
- Outlet Boundary Condition with a static pressure of 0 (zero) [Pa] at nine burner ports was defined.
- Symmetry Boundary Condition with geometric symmetry about the right vertical plane of the wind box was considered.

3. VARIOUS FLOW CONDITIONS

To estimate and to balance the secondary airflow across nine burners, various designs of the geometry have been proposed for the numerical simulation of the problem.

Full opening of the burner ports have been considered for balancing of the secondary airflow with baffle at two positions. Height of each groove is 1.6 inches.

When the ports were half open, the height of the grooves in the z-direction was changed to 0.8 inches. The dimensions of the baffle remained the same.

4. RESULTS AND DISCUSSION

In this section, the mass flow rate distribution across the 9 outlets is described. Numerical simulations with inlet mass flow rates of 0.178, 0.089, 0.355 Kg/ s were carried out using CFX5.6. All Numerical Simulations were solved at 2.6 million volume elements using the CFX Solver, which is the upper limit of the current computational facility.

4.1. FULL OPENING CASE WITH BAFFLE AT FIRST POSITION

Table 1. Mass flow rate (MFR) distribution at each burner under various flow conditions (ports are fully open and baffle at the first position)

Location	MFR (Kg/s)	fraction of MFR	½ MFR (Kg/s)	fraction of MFR	2 MFR (Kg/s)	fraction of MFR
Inlet	0.178		0.089		0.356	
Burner1	0.019	0.106	0.009	0.105	0.038	0.106
Burner2	0.020	0.114	0.010	0.116	0.038	0.106
Burner3	0.021	0.118	0.011	0.122	0.043	0.121
Burner4	0.018	0.102	0.009	0.100	0.037	0.103
Burner5	0.020	0.110	0.010	0.112	0.040	0.112
Burner6	0.022	0.122	0.011	0.122	0.044	0.125
Burner7	0.019	0.105	0.009	0.103	0.036	0.102
Burner8	0.019	0.108	0.009	0.104	0.036	0.100
Burner9	0.021	0.116	0.010	0.117	0.041	0.114
Sum	0.178	1.000	0.089	1.000	0.353	0.991

Table 1 shows the results obtained at a very refined mesh as mentioned above with MFR calculated at each burner.

From Table 1, it can be observed that mass flow rate defined at the inlet was recovered at the outlets, indicating the conservation of mass. The value of inlet MFR was determined from the scale model tests [1]. Figure 3, 4 and 5 represent a plot with burner location on x-axis and percentage of mass flow rate on y-axis is shown. The three figures show similar trend in the MFR distribution: the 3rd, 6th, and 9th burner receive more secondary airflow than others. This is due to the fact that these three burners are near the symmetry plane, and the airflow from both inlets contributes to their mass flow. Overall the balancing of secondary airflow at each burner can be observed when the ports are fully open for all three MFRs with the baffle at the first position, this suggests that the inlet MFR has little impact on the mass flow distribution across the burners when the ports are fully open and the baffle is at the first position.

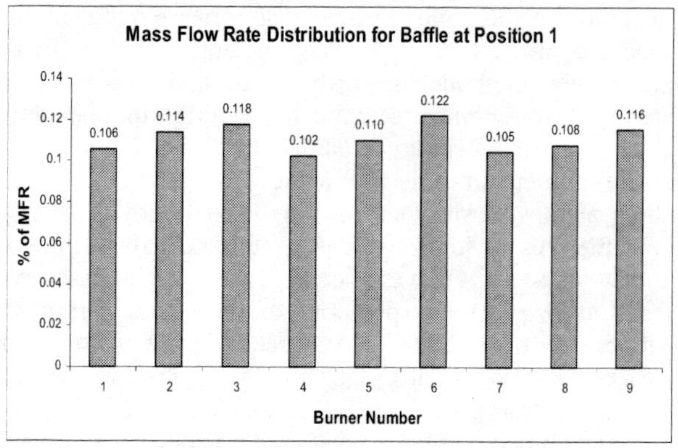

Figure 3. MFR distributions across each burner with inlet MFR = 0.177 Kg/s (ports are fully open, baffle at 1st position).

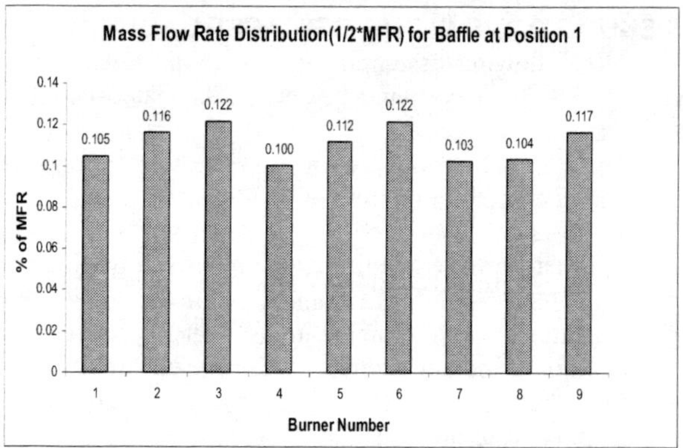

Figure 4. MFR distributions across each burner with inlet MFR = 0.089 Kg/s (ports are fully open, baffle at 1st position).

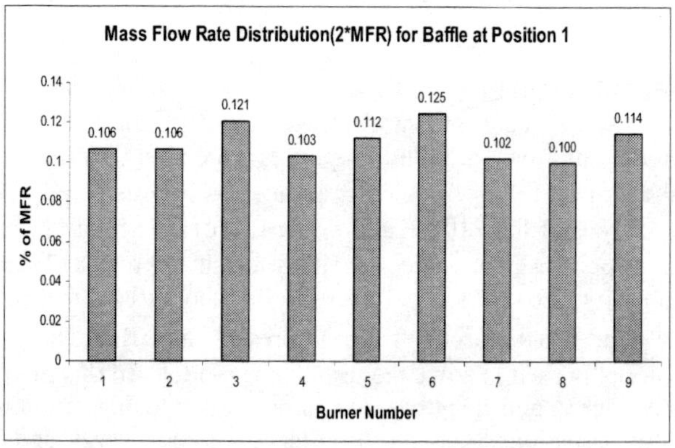

Figure 5. MFR distributions across each burner with inlet MFR = 0.356 Kg/s (ports are fully open, baffle at 1st position).

4.2. FULL OPENING WITH BAFFLE AT SECOND POSITION

Similarly for baffle at second position all three cases were considered and the plots are shown in Figure 6, 7, and 8. Although similar observations can be made regarding the mass flow rate distribution across each burner for the case when the baffle is at the second position, the uniformity of the MFR distribution appears not as good as in Figure 3, 4, and 5. The position of the baffle does have an influence on the MFR distribution. It should be taken into consideration in the wind box design.

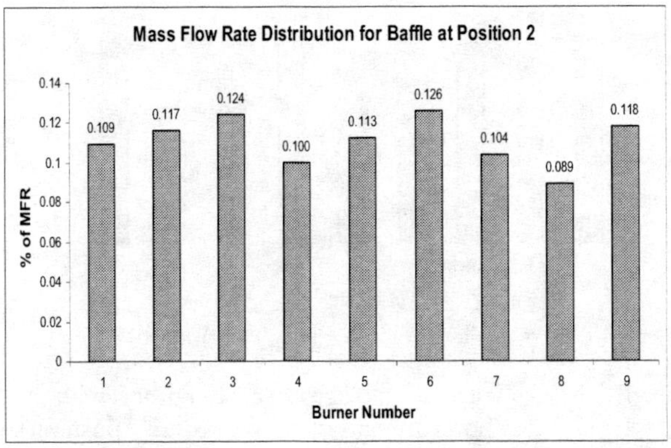

Figure 6. MFR distributions across each burner with inlet MFR = 0.177 Kg/s (ports are fully open, baffle at 2nd position).

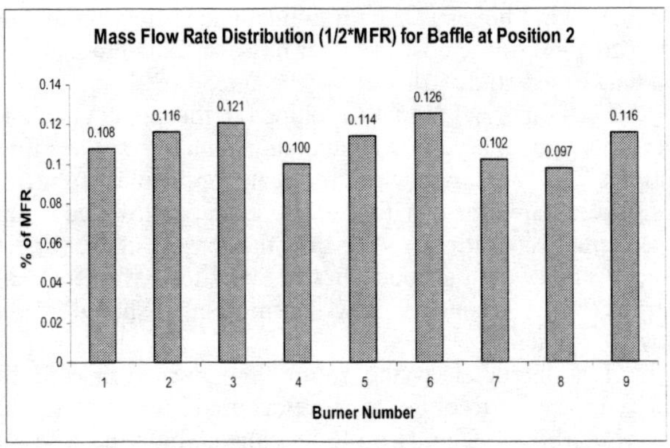

Figure 7. MFR distributions across each burner with inlet MFR = 0.089 Kg/s (ports are fully open, baffle at 2nd position).

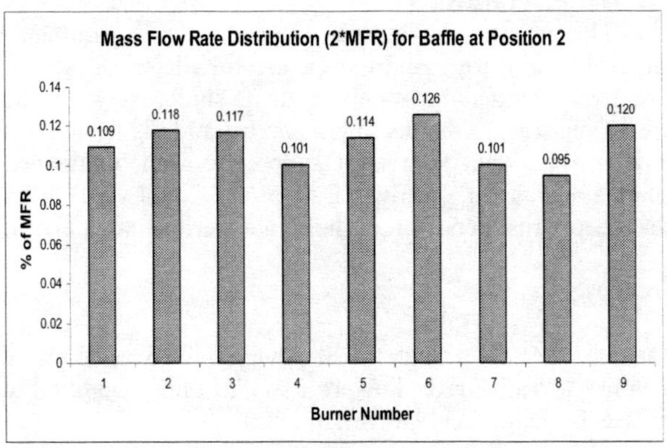

Figure 8. MFR distributions across each burner with inlet MFR = 0.356 Kg/s (ports are fully open, baffle at 2nd position).

In order to quantify the secondary flow balancing, standard deviations are tabulated for all the full open cases with baffle at two locations. A small value of the standard deviation indicates good balancing. From Table 2, when all the cases are compared, the second baffle position always give worse balancing compared to the first baffle position. Therefore, the second baffle position is not recommended.

Table 2. Comparison of standard deviation for full open cases

MFR (Kg/s)	For Full open Baffle at position1	For Full open Baffle at position2
	Standard Deviation	
MFR	0.0012	0.0021
1/2*MFR	0.0008	0.0009
2*MFR	0.0030	0.0037

4.3. HALF OPEN CASE WITH BAFFLE AT FIRST POSITION

Table 3 shows the results obtained at half open case with the same mass flow rates as in the scale model test. We can observe that when the ports are only half open, the balancing of the secondary flow becomes much worse than that for the fully open case. It shows that in terms of the secondary flow balancing, the current wind box and burner configuration does not perform very well under this particular inlet flow rate.

Figure 9 is the graph representing the comparison of percentage of MFR values obtained numerically and experimentally. There are discrepancies in the results, especially at burner number 1. The reason for these discrepancies is believed to be due to inlet boundary condition imposed in the numerical simulation. In CFX5.6, an imposed mass flow rate at inlet implies uniform velocity distribution. However, in the scale model test, flow enters the wind box through a long curved duct with a rectangular cross section, which results in quite a quite non-uniform velocity distribution at the wind box inlet. Although the mass flow rate is reproduced in the CFD modeling, the detailed non-uniform velocity distribution cannot be exactly reproduced in the numerical modeling. Apparently, the velocity distribution at the wind box inlet affects significantly the balancing of the secondary flow. Burner number 1 is the closest one to the inlet and the largest discrepancy between the simulation and measurement is reported. It is intended in the future study, the flow in the duct will be included so that in the simulation a more realistic flow at the wind box inlet can be obtained.

Table 3. Comparison the mass flow rate (MFR) distribution at each burner (ports are half open and baffle at first position) obtained from the numerical simulation and the scale model measurements

Location	Numerical Analysis		Experimental Measurements	
	MFR Kg/s	fraction of MFR	MFR Kg/s	fraction of MFR
Inlet	1.212		1.212	
Burner1	0.139	0.114	0.120	0.099
Burner2	0.142	0.117	0.131	0.108
Burner3	0.139	0.114	0.140	0.116
Burner4	0.127	0.105	0.119	0.098
Burner5	0.127	0.105	0.138	0.114
Burner6	0.141	0.117	0.133	0.109
Burner7	0.132	0.109	0.128	0.106
Burner8	0.126	0.104	0.132	0.109
Burner9	0.138	0.114	0.133	0.110
Sum	1.212	1.000	1.175	0.969

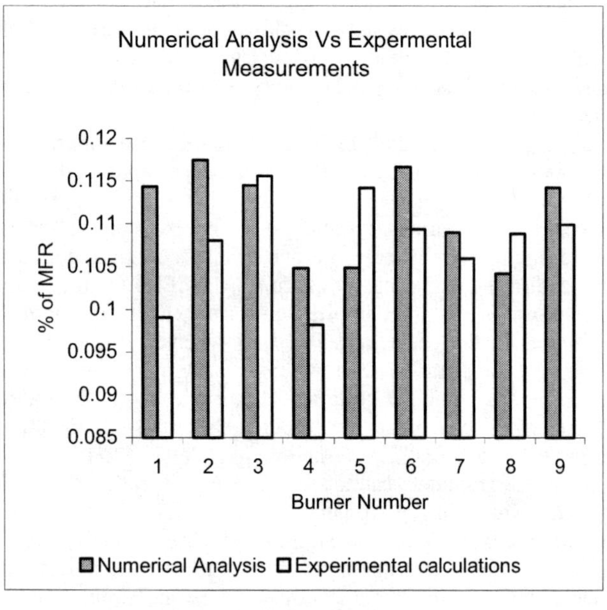

Figure 9. Graph representing the comparison of percentage of MFR values obtained numerically and experimentally (ports are half open and baffle at 1st position).

4.4. HALF OPEN CASE WITH BAFFLE AT SECOND POSITION

As it is possible in numerical analysis to check the balancing of the secondary airflow with the change in baffle position, the baffle has therefore been moved to the second position mentioned. Graphical representations of the values obtained numerically are shown in Figure 10 with baffle at second position.

Figure 10 shows the second position of the baffle provides detrimental effects on the secondary flow balancing across the burners. It should be avoided in the design.

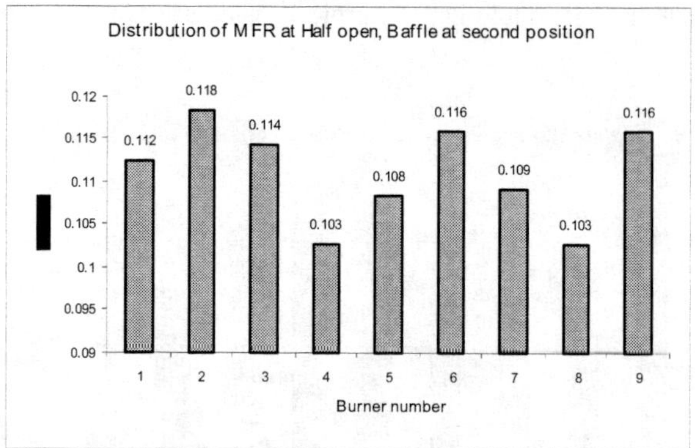

Figure 10. MFR distributions across each burner with inlet MFR = 1.212 Kg/s (ports are half open, baffle at 2nd position).

5. CONCLUSIONS

This paper presents methodology used to numerically model a scaled power plant wind box. Numerical modeling of power plant wind boxes is shown to be a viable flow prediction tool at a fraction of the cost of either physical flow modeling or a field data based study [3].

The numerical model developed in this paper is able to predict the secondary flow balancing under various flow conditions. The comparison with the scale model measurements shows general agreement with the discrepancy attributed to the imposed inlet condition. For all the cases, it has been concluded that the second baffle position worsens the uniformity of the secondary flow distribution. It should not be employed in practice.

It is intended in the future, the duct flow will be considered in order to obtain a more accurately inlet condition at the wind box. The two-phase calculation will also be performed to identify the optimum distribution of the dust deposition on the wind box floor.

ACKNOWLEDGMENTS

The first author would like to thank the Department of Mechanical Engineering and the Centre for Electric Power for their support provided during his graduate study. He would like to express his sincere thanks to his parents and his friends who have been a constant source of support and encouragement. Graduate students Mr. Srinivas Badam and Mr. Evan Ballard, are also deeply thanked for their help in experimental analysis.

References

[1] Malone, G.B. "Computational Analysis of Secondary Air Flow in a Wall-Fired Power Plant", PhD Dissertation, Tennessee Technological University, 2002

[2] CFX, Version 5.6, Copyright 1996-2003 by ANSYS Inc., Pennsylvania, USA.

[3] LaRose, J.A., Hopkins, M.W. "Numerical Flow Modeling of Power plant Windboxes," Power-Gen Americas '95, Barberton, Ohio, USA.

[4] Sellers, K. "Secondary Air Flow Balancing in Wall-Fired Units," M.S. Thesis, Tennessee Technological University, 2000.

[5] Armfiled, S., Fletcher, C. "Comparison of k-e and Algebraic Reynolds Stress Models for Swirling Diffuser Flow," International Journal for Numerical Methods in Fluids, vol. 9, pp. 987-1009, 1989.

[6] Ferziger, J., Peric, M. Computational Methods for Fluid Dynamics, 2nd edition, Springer, 1999.

[7] Wilcox, D. Turbulence Modeling for CFD, CDW Industries, November 1994.

RECENT ADVANCES IN SOLIDS AND STRUCTURES

INTRODUCTION

This collection represents those papers presented at the symposium entitled, "Recent Advances in Solids and Structures," as a part of the program at the 2004 International Mechanical Engineering Congress, held in Anaheim, California. This symposium is celebrates its 14TH year.

The papers included in this collection cover various subjects in solids and structures ranging from traditional areas to nanotechnology. Some of them are academic problems while others are practical industrial problems. The topics include crack, fatigue, vibration and control, composite structures, etc. The papers highlight state of the arts technology and research in these subject areas. The editors hope this symposium volume will serve as a useful reference for researchers and practitioners in the subject areas.

The editors wish to express sincere appreciation to the authors of this publication.

Howard H. Chung
MITEC International, Inc.
Naperville, Illinois

Young W. Kwon
Southern Illinois University
Carbondale, Illinois

Proceedings of IMECE04
2004 ASME International Mechanical Engineering Congress and Exposition
November 13-20, 2004, Anaheim, California USA

IMECE2004-59399

MOLECULAR DYNAMICS STUDY OF METAL FATIGUE PROCESS

Y. W. Kwon
Mechanical Engineering & Energy Processes Department
Southern Illinois University
Carbondale, IL

ABSTRACT

Molecular dynamics study was conducted to understand fatigue process in metals and to predict fatigue failure. As the first step, a pure metal like copper was considered for the study with defects at the atomic level such as vacancies or dislocations. The study was focused on identifying parameters which can provide indications of progressive damage accumulation in the material under cyclic loading. The results obtained by simulations were compared to macroscopic observations in the experimental studies

INTRODUCTION

Fatigue is damage-accumulation process of real materials under cyclic loads. Defects, such as impunities, inclusions, vacancies, and dislocations, play an important role in fatigue. When a material including those defects is stressed even in the elastic range, there are irreversible changes within the material, which are initially so small that such changes are not measurable in the stress-strain curve. However, as the number of cycles in the loads increases significantly, those changes lead to ultimate failure of the materials [1].

So far, design of materials or structures has been heavily relied on experimental data, which were obtained through expensive and very time-consuming processes. There are so many parameters that affect fatigue behavior. Some of them are random stress fluctuations, stress concentrations, surface roughness, residual stress, corrosive environment, salinity, etc. Considering all possible combinations of those factors in fatigue experiments is almost impossible and prohibitive in terms of the practical point of view. Furthermore, as nanotechnology has been pushed forward recently, many devices have been developed in a very small scale such as nano- or at most micro-scale. Even these scale specimens were shown to fail as a result of fatigue [2]. In these scales, experimental testing for fatigue behavior is quite a challenge.

Analytical modeling and simulation of fatigue process is very beneficial to alleviate the heavy burden of experimental work as well as to enhance the understanding of fatigue process. The present work is to utilize the molecular dynamics to understand the fatigue damage mechanism in metallic materials.

Molecular dynamics has been applied to predict material behaviors [3-5] and fracture process with existing cracks. Much of the research was focused on nano-structures such as nanotubes [6-9]. Another group of efforts were put into modeling and understanding crack behavior and fractures [10-21]. On the other hand, very limited efforts were spent for modeling and simulation of fatigue process as discussed below.

Chang and his coworker [22, 23] conducted studies for the tension-compression fatigue test for nanoscale coppers with vacancies at different temperatures using the Lennard-Jones potential. The studies plotted the fatigue strength vs. number of cycles to failure, *S-N* curve. The simulation considered about 960 atoms and up to 100,000 cycles. From the studies, it was concluded that a lower vacancy fraction (= number of vacancies / total number of atoms) or a higher temperature yielded a higher endurance limit in the *S-N* curve, and a low stress caused ductile fracture in fatigue failure. Despite the findings, those studies did not address the fatigue mechanisms to understand the fatigue damage process. Furthermore, the Lennard-Jones potential is more suitable for amorphous materials rather than for crystalline solids.

Yasukawa [24] studied the influence of moisture on the static-fatigue strength of silicon oxide using extended Tersoff interatomic potential. The potential considered interatomic charge transfers between atoms of different species during the process of their reaction. In this study, the author explained the mechanism of strength decrease of silicon oxide due to moisture.

Ye, Farkas and Soboyejo [25] studied fracture toughness and fatigue crack growth experimentally and theoretically. They used the embedded atom method [26, 27] for the theoretical study, which was used to explain some experimental findings. In this study, they focused on understanding the reasons for improved toughness of 40Ti alloy with an incrementally increasing load, but not a cyclic load. Fatigue studies mentioned above only focused on crack behaviors of fatigue damage process.

The objective of this study was to understand the fatigue damage process and to identify a parameter(s) that could be correlated to the fatigue life, i.e. the number of cycles to failure. In order to achieve the objective, the embedded atoms method [26, 27] was adopted in this study.

ANALYSIS & RESULTS

In the embedded atom method [26, 27], the energy of a collection of atoms is determined by the electron density, i.e. the distribution of electrons in the system. As a result, the energy consists of two parts: the energy to embed each atom into the electron density of the neighboring atoms, and the energy by a short range pair interaction for the core-core repulsion. The energy of a system can be represented by

$$E_{tot} = \sum_i E_{ee}(\rho_{h,i}) + \frac{1}{2}\sum_{i,j}\phi_{ij}(r_{ij}) \qquad (1)$$

where $\rho_{h,i}$ is the total electron density seen by atom i due to the surrounding atoms in the system, E_{ee} is the embedding energy for placing an atom into that electron density and ϕ_{ij} is the short range pair interaction for the core-core repulsion between atoms i and j separated by distance r_{ij}.

In Eq. (1), the total electron density seen by an atom is computed by summation of atomic densities in surrounding atoms. That is,

$$\rho_{h,i} = \sum_{j \neq i} \rho_j^a(r_{ij}) \qquad (2)$$

in which ρ_j^a is the electron density contributed by atom j with distance r_{ij} from atom i. The short range, repulsive pair potential used in Ref. [26, 27] is also used and it is expressed as

$$\phi_{ij}(r) = Z_i(r)Z_j(r)/r \qquad (3)$$

The function $Z(r)$ is determined by fitting to the universal binding function.

Pure copper was selected in this study. Because copper is a Face-Centered Cubic (FCC) crystal structure, a three-dimensional arrangement of 490 atoms with FCC crystal structures is shown in Fig. 1. The figure shows the *yz*-plane view with vacancies that extends along the *x*-axis. The plotted atomic arrangement is before equilibrium of the atoms among themselves.

Classical Molecular Dynamics (MD) was applied to the system under an isothermal condition, i.e. at a constant temperature. The Embedded Atom Method (EAM) was used to model the copper atoms. First of all, MD was applied to the atoms without external loading to determine the equilibrium configuration, and then later MD was again applied to the equilibrium configuration with cyclic loads. The atoms on the left side plane (negative *xz*-plane) were constrained while the atoms on the right side plane (positive *xz*-plane) were subjected to tensile cyclic loading of the sinusoidal shape along the *y*-axis. The average potential energy per atom in the copper system is plotted in Figure 2 as a function of the number of cycles. As shown in the figure, the potential energy increases along with the number of cyclic loads. When a linear curve fit is made to the potential energy variation, the slope represents the rate of potential energy increase (i.e. the change of potential energy per load cycle). In order to determine the fatigue life, the separation energy of the atoms is divided by the slope.

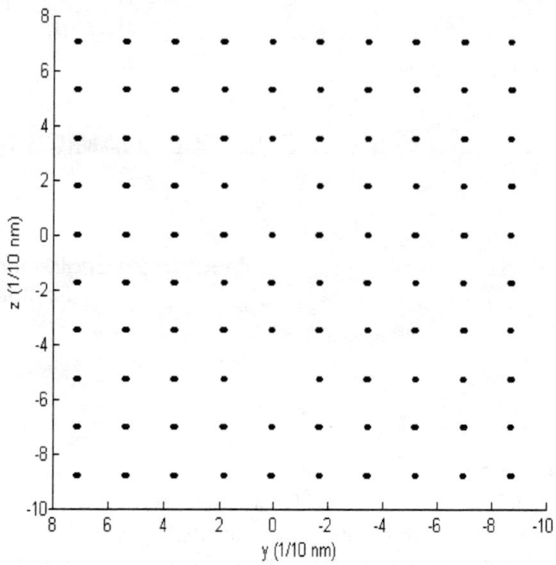

Figure 1. Plane view of 3-D atomic arrangement with two rows of vacancies

Increase of potential energy can be explained in the following way. For mathematical simplicity, consider a simple spring and mass system as shown in Figure 3. If the spring is linear, there is resonance when the excitation frequency matches the natural frequency of the system. This results in the potential energy increase in the system. In MD simulation, the spring system is nonlinear, i.e. a function of displacement. Even in this case, as the excitation frequency gets closer to the natural frequency, the potential energy increases.

In addition, random velocity distribution among atoms can cause instability at the atoms with high velocities. Let us consider an interaction between two atoms whose interaction is described in terms of a nonlinear potential function. One of the two atoms is fixed while the other vibrates. Assuming one-dimensional motion, the dynamic equation of the vibrating atom can be written as

$$m\ddot{x} + f(x) = 0 \qquad (4)$$

in which m is the atomic mass, x is the displacement of the vibrating atom from its equilibrium position, and $f(x)$ is the interaction force between the atoms, which is derived from the interaction potential. Rewriting the equation in terms of state variables, displacement $x_1 = x$ and velocity $x_2 = \dot{x}$, yields

$$\begin{aligned}\dot{x}_1 &= x_2 \\ \dot{x}_2 &= -f(x_1)/m\end{aligned} \qquad (5)$$

The plot of phase trajectory of Eq. (5) indicates the stability of the system and it depends on the initial conditions of the displacement and velocity.

Velocities of atoms under equilibrium obey Maxwell-Boltzmann law. Considering one-dimensional problem, the fraction of atoms having velocities between v and $v+dv$ can be expressed as

$$g(v)dv = \sqrt{\frac{m}{2\pi kT}} \exp\left(-\frac{mv^2}{2kT}\right)dv \qquad (6)$$

where k is Boltzmann's constant and T is the temperature. Therefore, some of the atoms have much higher velocities than the average velocity of the system. Those atoms with high velocities show unstable phase trajectory as shown in Fig. 4 while atoms with low velocities have stable phase trajectory as shown in Fig. 5. The unstable phase trajectory means increase of position or increase of potential energy. In other words, when a system of atoms is subjected to an external loading, some atoms at unstable equilibrium will increase potential energy, thus contributing to the increase of the total potential energy in the system.

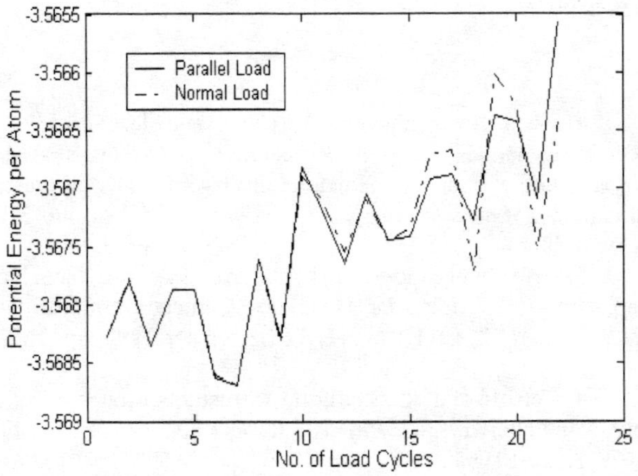

Figure 2. Plot of average potential energy per atom in the system with vacancies vs. the number of cyclic loads at the unloaded state

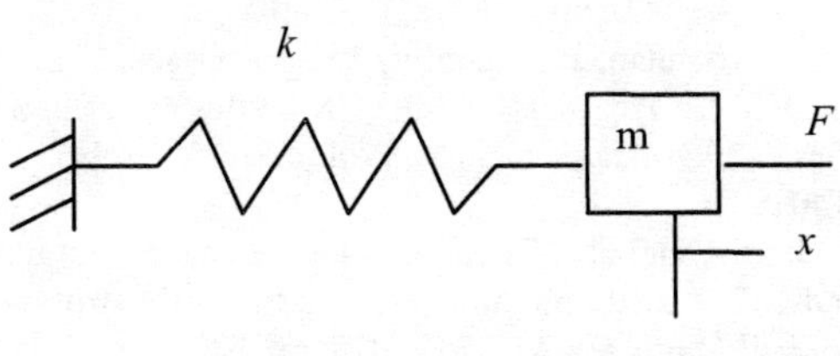

Figure 3. One-dimensional linear spring-mass system subjected to a force

The next example cases considered the effects of mean stress on the fatigue life. In this model, a dislocation was introduced into the system instead of vacancies. In these cases, the stress amplitudes (σ_a) of the cyclic loading remained the same but the mean stress (σ_m) of the cyclic loading varied. Previous experimental study showed that the increase of mean stress resulted in a shorter fatigue life [29]. Figure 6 compares the increases of normalized potential energy per atom for two different ratios of mean stress to stress amplitude, which were obtained again through linear curve-fitting of the potential energy. The first case has the ratio of mean stress to the stress amplitude of cyclic loads equal to $\sigma_m/\sigma_a=1$ while the other case has the ratio of $\sigma_m/\sigma_a=2$. In both cases, the stress amplitudes were held the same. The higher mean stress case yielded a shorter fatigue life as shown in Fig. 4. The normalized fatigue life prediction from the present study is plotted for different ratios of σ_m/σ_a as seen in Fig. 7. As shown in the figure, the fatigue life decreased almost linearly in the given range of the stress ratios σ_m/σ_a.

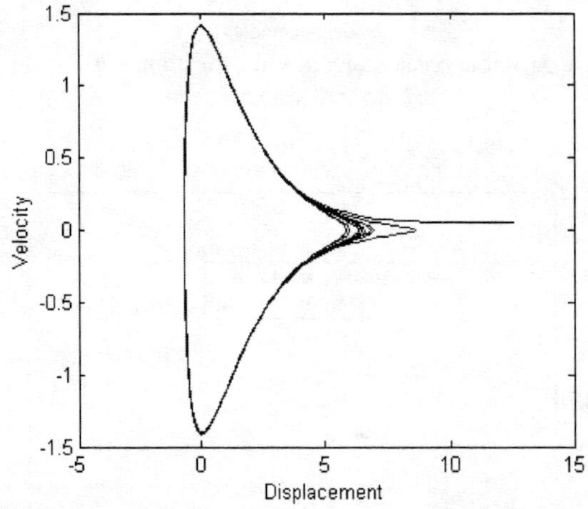

Figure 4. Unstable phase trajectory for the atom with an initial high velocity

CONCLUSIONS

Molecular dynamics simulation was conducted to predict the fatigue damage process in pure copper. Pure copper with defects was considered in the model with cyclic loads. The study showed that increase of potential energy in the system would be an indicator for fatigue damage accumulation. The predicted results of fatigue life with different mean stresses agreed well with previous experimental data.

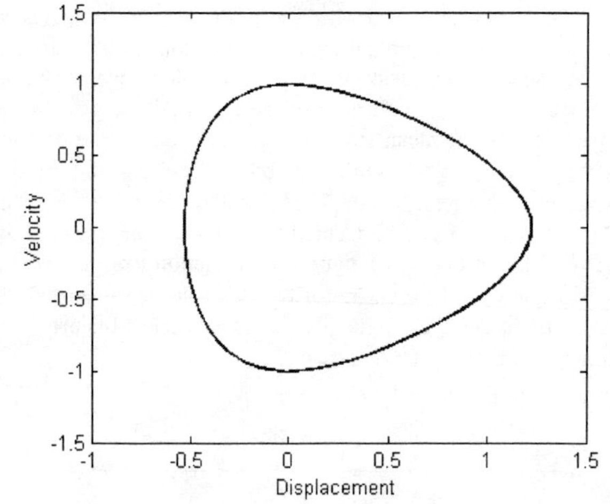

Figure 5. Stable phase trajectory for the atom with an initial normal velocity

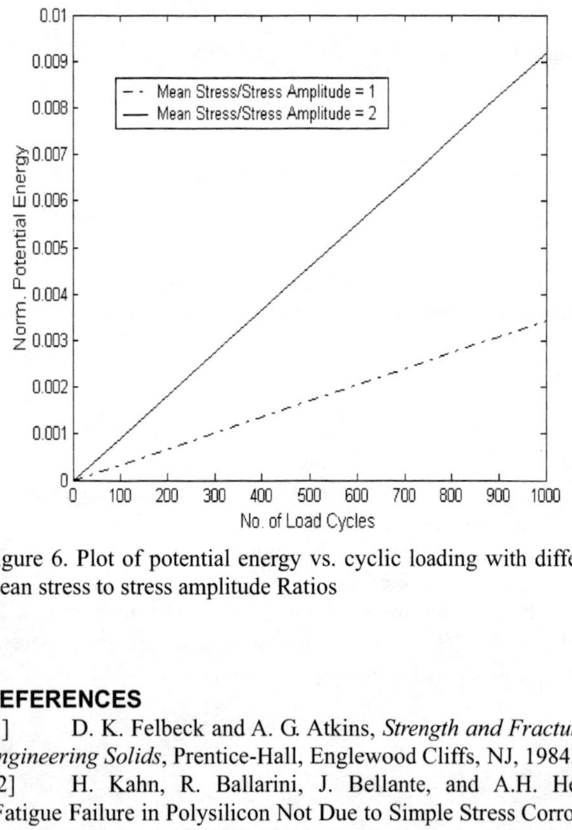

Figure 6. Plot of potential energy vs. cyclic loading with different mean stress to stress amplitude Ratios

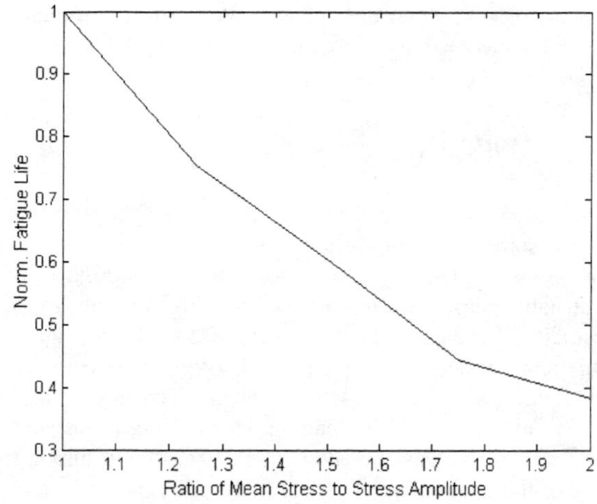

Figure 7. Predicted fatigue life reduction resulted from increase of the mean stress

REFERENCES

[1] D. K. Felbeck and A. G. Atkins, *Strength and Fracture of Engineering Solids*, Prentice-Hall, Englewood Cliffs, NJ, 1984.

[2] H. Kahn, R. Ballarini, J. Bellante, and A.H. Heuer, "Fatigue Failure in Polysilicon Not Due to Simple Stress Corrosion Cracking," *Science*, Vol. 298, November 8, 2002, pp. 1215-1218.

[3] A. M. Krivtsov and M. Wiercigroch, "Molecular Dynamics Simulation of Mechanical Properties for Polycrystal materials", *Mater. Phys. Mech*, Vol. 3, 2001, pp.45-51.

[4] H. van Swygenhoven, A. Caro, and D. Farkas, "A Molecular Dynamics Study of Polystalline FCC Metals at the Nanoscale: Grain Boundary Structure and Its Influence on Plastic Deformation", *Materials Science and Engineering A*, Vol. 390-310, 2001, pp. 440-444.

[5] H. van Swygenhoven and A. Caro, "Plastic Behavior of Nanophase Ni: A Molecular Dynamics Computer Simulation", *Appl. Phys. Lett.*, Vol. 71, No. 12, 22 September 1997, pp. 1652-1654.

[6] J. Tersoff and R. S. Ruoff, "Structural Properties of a Carbon-Nanotube Crystal", *Physical Review Letters*, Vol. 73, pp. 676-679, August 1994.

[7] C. F. Cornwell and L. T. Wille "Elastic Properties of Single-Walled Carbon Nanotubes in Compression", *Solid State Communications*, Vol. 101, 1997, pp. 555-558.

[8] T. Ozaki, Y. Iwasa, and T. Mitani, "Stiffness of Single-Walled Carbon Nanotubes under Large Strain", *Physical Review Letters*, Vol. 84, 2000, pp. 1712-1715.

[9] A. Krishnan, E. Dujardin, T. W. Ebbesen, P. N. Yianilos, and M. M. J. Treacy, "Young's Modulus of Single-Walled Nanotubes", *Physical Review B*, Vol. 58, November 1998, pp. 14013-14019.

[10] J. E. Sinclair, "The Influence of the Interatomic Force Law and of Kinks on the Propagation of Brittle Crack", *Philosophical Magazine*, Vol. 31, 1975, pp. 647-671.

[11] S. Kohlhoff, P. Gumbsch, and H. F. Fischmeister, "Crack Propagation in B.C.C. Crystals Studied with a Combined Finite-Element and Atomistic Model", *Philosophical Magazine A*, Vol. 64, 1991, pp. 851-878.

[12] M. S. Daw, M. I. Baskes, C. L. Bisson, and W. G. Wolfer, "Application of the Embedded Atom Method to Fracture, Dislocation Dynamics, and Hydrogen Embrittlement", Sandia Report, SAND86-8660, March 1986.

[13] C. Hsieh, R. Thomson, "Lattice Theory of Fracture and Crack Creep", *Journal of Applied Physics*, Vol. 44, 1973, pp. 2051-2063.
[14] R. G. Hoagland, M. S. Daw, S. M. Foiles, and M. I. Baskes, "An Atomic Model of Crack Tip Deformation in Aluminium Using an Embedded Atom Potential", *J. Mater. Res.*, Vol. 5, 1990, pp. 313-324.
[15] M. Mullins, "Computer Simulation of Fracture Using Long Range Pair Potentials", *Acta Metall.*, Vol. 32, 1984, pp. 381-388.
[16] Y. W. Zhang and C. T. Wang, "Lattice Instability at a fast Moving Crack", *J. Appl. Phys.*, Vol. 80, No. 8, 15 October 1996, pp.4332-4336.
[17] Y. W. Zhang, C. T. Wang, and Q. H. Tang, "Simulation of nucleation and Emission of Dislocations by Molecular-Dynamics Method", *J. Appl. Phys.*, Vol. 77, No. 6, 15 March 1995, pp. 2393-2399.
[18] K. S. Cheung, A. S. Argon, and S. Yip, "Activation Analysis of Dislocation Nucleation from Crack Tip in α-Fe", *J. Appl. Phys.*, Vol. 69, No. 4, 15 February 1991, pp. 2088-2096.
[19] S. Blonski and W. Brostow, "Molecular Dynamics Simulations of Chain relaxation and Crack propagation in Polymer Liquid Crystals", *J. Chem. Phys.*, Vol. 95, No. 4, 15 August 1991, pp. 2890-2896.
[20] Z. Li, W. Chu, J. Li, K. Gao and L. Qiao, "Three-Dimensional Molecular Dynamics Simulation of Hydrogen-Enhanced Dislocation Emission and Crack propagation", *Progress in Natural Science*, Vol. 12, No. 9, September 2002, pp. 717-720.
[21] P. Walsh, R. K. Kalia, A. Nakano, and P. Vashishta, "Amorphization and Anisotropic Fracture Dynamics during Nanoindentation of Silicon Nitride: A Multimillion Atom Molecular Dynamics Study", *Appl. Phys. Lett.*, Vol. 77, No. 26, 25 December 2000, pp. 4332-4334.
[22] W.-J. Chang, "Molecular-Dynamics Study of Mechanical Properties of Nanoscle Copper with Vacancies under Static and Cyclic Loading", *Microelectronic Engineering*, Vol. 65, 2003, pp. 239-246.
[23] W.-J. Chang and T.-H. Fang, "Influence of Temperature on Tensile and Fatigue Behavior of Nanoscale Copper Using Molecular Dynamics Simulation", *Jour. of Physics and Chemistry of Solids*, Vol. 64, 2003, pp. 1279-1283.
[24] A. Yasukawa, "Using an Extended Tersoff Interatomic Potential to Analyze the Static-Fatigue Strength of SiO_2 under Atmospheric Influence", *JSME International Journal*, Series A, Vol. 39, No. 3, 1996, pp. 313-320.
[25] F. Ye, D. Farkas, and W. O. Soboyejo, "An Investigation of Fracture and Fatigue Crack Growth Behavior of Cast Niobium Aluminide Intermetallics", *Materials Science and Engineering A*, Vol. 264, 1999, pp. 81-93.
[26] M. S. Daw, and M. I. Baskes, "Embedded Atom Method: Derivation and Application to Impurities, Surfaces, and Other Defects in Metals," *Physical Review B,* v. 29, n. 12, 15 June 1984, pp. 6443-6453.
[27] M. S. Daw, "Model of Metallic Cohesion: the Embedded-Atom Method," *Physical Review B,* v. 39, n. 11, 15 April 1989-I, pp. 7441-7452.
[28] M. A. Miner, "Comulative Damage in Fatigue"J. Appl. Mech. Vol. 12, Trans. Am. Soc. Mech. Engrs, Vol. 67, 1945, pp. A159-A164.
[29] N. E. Frost, K. J. Marsh, and L. P. Pook, *Metal Fatigue*, Clarendon Press, Oxford, 1974.

Proceedings of IMECE04
2004 ASME International Mechanical Engineering Congress and Exposition
November 13-20, 2004, Anaheim, California USA

IMECE2004-59433

PASSIVE VIBRATION DAMPING OF ALUMINUM BY 1-3 PIEZOCOMPOSITES

A. Saigal, R. Greif and S. Nakhwa
Department of Mechanical Engineering
Tufts University
Medford, MA 02155

ABSTRACT

The effective properties of 1-3 piezocomposites are used to examine their passive vibration damping characteristics. An aluminum cantilever beam bonded with 1-3 piezocomposite dampers is modeled by means of "ANSYS" and "SIMULINK" softwares to investigate the dynamic behavior of the system. A method of determining the damping ratio introduced by the piezocomposite damper in conjunction with a simple resistive electrical circuit is established. The effect of volume fraction of the 1-3 piezocomposite on the damping of the system is analyzed. Damping ratio is observed to increase with rising volume fraction. At low volume fractions, the participation of piezoelectric fibers in the load-bearing pattern is to a lesser extent and hence the damping ratio is low. On the contrary as the volume fraction rises, the involvement of piezoelectric fibers increases resulting in higher damping ratios. Given that the inherent material damping in the aluminum beam is 0.0002, the additional damping provided by the bonded piezoelectric strips goes up to a maximum of 0.0042. Finally, the methodology developed in this paper can be used to model any type of vibratory structural system to determine the damping introduced by the piezocomposites.

INTRODUCTION

Damping is required in structural systems to decrease the peak vibration amplitudes and avoid resonance, thereby adding robustness and stability to the system. Piezoelectric materials, because of their peculiar and inherent energy transformation properties, have tremendous potential in the vibrations field. According to Forward et al. [1], Crawley et al. [2] and Hagood et al. [3], they are used as structural actuators, sensors and dampers. Damping provided by piezoelectric materials is of two types: active damping and passive damping.

In active damping the piezoelectric damper is used as an actuator and a sensor, which requires costly and sophisticated amplifiers and control electronic equipment. These can be eliminated in passive damping applications where the only external element is a simple passive electrical circuit. Modeling of damping, with piezoelectric materials and passive electrical circuits, is described in by Hagood et al. [4].

This study concentrates on analysis of damping of a cantilever beam by means of a piezoelectric device. Popular practice and best arrangement for maximizing damping is to bond two piezocomposites on top and bottom near the fixed end of the beam. These locations have peak strains and are known to give the utmost structural damping [5]. Figure 1 shows the cantilever beam with bonded piezocomposite elements. Ignoring the other forms of material and structural damping inherent in the system, the impact of just the piezoelectric material on the system's dynamic performance is examined.

Material Properties

The cantilever beam is made of aluminum 12"x1"x1/8" in dimensions. Two piezocomposite strips of 2"x1"x0.01" are bonded on the top and bottom of the beam near the fixed end. Properties of the aluminum [6] cantilever beam material are listed in Table 1.

Table 1: Aluminum Data

Material Property	Value
Young's Modulus	70 GPa
Density	2778 kg/m^3
Poisson's Ratio	0.3

MATERIAL PERSPECTIVE

The constitutive equation utilized in the passive damping analysis in matrix form is given by [8]:

$$\begin{bmatrix} \sigma \\ D \end{bmatrix} = \begin{bmatrix} C^E & -e^T \\ e & k^\varepsilon \end{bmatrix} \begin{bmatrix} \varepsilon \\ E \end{bmatrix} \tag{1}$$

where σ is the stress tensor, ε is the strain tensor, D is the dielectric displacement vector and E is the electric field vector. C^E represents the stiffness tensor, k^ε is the electric permittivity tensor and e^T is the piezoelectric coupling tensor, where the superscripts E and ε indicate constant electrical field and strain boundary conditions respectively, and T denotes matrix transpose.

The effective elastic, piezoelectric and dielectric properties of the 1-3 piezocomposite dampers consisting of PZT-7A fibers arranged hexagonally in an epoxy matrix are determined by the "ANSYS" [7] finite element analysis and are listed in Table 2 [9].

Table 2: Effective Bulk Properties of PZT-7A and Epoxy 1-3 Piezocomposite.

	Hex Topology PZT-7A Composite Effective Properties										
Vol. Fract.	C_{11} GPa	C_{12} GPa	C_{13} GPa	C_{33} GPa	C_{44} GPa	C_{66} GPa	e_{31} C/m^2	e_{33} C/m^2	e_{51} C/m^2	k_{11} nC/Vm	k_{33} nC/Vm
15%	9.781	5.218	5.3	18.916	2.34	2.28	-0.024	1.595	0.00259	0.05	0.341
25%	11.32	5.93	6.08	26.313	2.798	2.698	-0.045	2.654	0.00489	0.061	0.544
30%	12.26	6.365	6.58	30.112	3.067	2.947	-0.057	3.184	0.00677	0.068	0.646
40%	14.52	7.349	7.68	37.715	3.703	3.588	-0.087	4.24	0.01200	0.085	0.849
50%	17.74	8.575	9.23	45.661	4.566	4.572	-0.127	5.287	0.02200	0.109	1.051
60%	22.26	10.36	11.4	54.062	5.714	5.968	-0.186	6.321	0.04100	0.144	1.255
100%	154.84	83.24	82.7	131.4	25.7	35.8	-2.12	9.52	9.35	4.065	2.079

Hagood et al. [4] presented a comprehensive analytical model of the shunted piezoelectric damper as shown in Figure 2. The piezoelectric is considered analogous to a transformer that converts input mechanical energy into electrical energy. Some of the electrical energy is stored in the capacitor and the rest is dissipated as heat by the resistor. The capacitor is not external but is the inherent capacitance of the piezoelectric itself because of its dielectric nature.

To simplify the analysis, uniaxial loading assumption is made which results in three modes of loading.

i) Longitudinal: Force and field in poling direction.
ii) Transverse: Field in poling direction and force in a direction perpendicular to it.
iii) Shear: Force in shear direction and field in a direction perpendicular to the poling direction, respectively.

For these cases, the expressions for electromechanical coupling coefficients k_{ij} are given by Mason [10]. "k_{ij}^2 is defined as the ratio of peak energy stored in the capacitor to the peak energy stored in mechanical strain under uni-axial loading and harmonic motion". Mathematically for force in the jth direction and field in the ith direction,

$$k_{ij} = \frac{d_{ij}}{\sqrt{s_{jj}^E \varepsilon_i^T}} \tag{2}$$

According to Hagood et al. [4], maximum loss factor η for a particular loading mode is given by

$$\eta_{\max jj}^{RES}(\omega) = \frac{k_{ij}^2}{2\sqrt{1-k_{ij}^2}} \tag{3}$$

where the RES superscript refers to resistor shunting.

In practical damping applications, the piezoelectric damper is coupled to the structure to be damped, wherein the loss factor is an average of the system component loss factors weighted by the fraction of strain energy in the respective elements [5].

$$\eta^{TOT} = \frac{\sum_{i=1}^{n} \eta_i U_i}{\sum_{i=1}^{n} U_i} \tag{4}$$

where U_i is the strain energy in the ith element of the structure. To take advantage of the weighting, the design criterion is to place the piezoelectric material in areas of high strain energy to maximize the structural damping.

The generalized electromechanical coupling coefficient K_{ij} for a piezoelectrically-damped cantilever beam is given by [4]

$$K_{ij}^2 = \left(\frac{\omega_n^{D^2} - \omega_n^{E^2}}{\omega_n^{E^2}} \right) \tag{5}$$

where ω_n^D and ω_n^E are the open circuit and short circuit natural frequencies of the system.

MODELING

"ANSYS" Contribution in Modeling

The ANSYS [7] finite element software is utilized to determine the piezocomposite open circuit (ω_n^D) and short circuit (ω_n^E) natural frequencies of the system. Open circuit being stiffer than the short circuit piezocomposite will make the structure vibrate at a higher natural frequency. Other useful information obtained from ANSYS, that is required by the SIMULINK [11] model is the peak strain energy in the aluminum beam, mechanical strain energy and total energy in the piezocomposite. The natural frequencies and the peak energies during vibration after release from the initial displacement of 3 mm for various volume fractions of 1-3 PZT-7A composites are calculated and shown in Table 3.

"SIMULINK" Contribution in Modeling

SIMULINK [11] is used to process further the ANSYS derived data of natural frequency and energies to arrive at the energy loss per cycle in the resistor. From this information, the overall loss factor of the system and damping added by the piezocomposite device is determined. It is also possible to plot the beam tip deflection with respect to time and find the logarithmic decrement of the amplitude. The block diagram of the system is shown in Figure 3.

The natural frequency of vibration of the system will be between the open circuit and short circuit frequencies. However, since the difference between the two frequencies is negligible, the short circuit frequency is taken as the system natural frequency of vibration.

Table 3: ANSYS results for the vibration model

			Energies in the system		
Vol. Fr. (%)	(ω_n^E) (Hz)	(ω_n^D) (Hz)	Al Strain Energy (J)	Piezo Total Energy (J)	Piezo Strain Energy (J)
15%	28.682	28.694	0.112414e-2	0.266881e-4	0.248413e-4
25%	28.752	28.766	0.112451e-2	0.311355e-4	0.291936e-4
30%	28.793	28.808	0.112474e-2	0.337045e-4	0.316687e-4
40%	28.890	28.907	0.112536e-2	0.397399e-4	0.374372e-4
50%	29.027	29.048	0.112641e-2	0.482413e-4	0.454503e-4
60%	29.209	29.237	0.112813e-2	0.593976e-4	0.557760e-4
Monoli.	32.067	32.359	0.121262e-2	0.204575e-3	0.164358e-3

According to Hagood et al. [4] the piezocomposite during resistive damping exhibits frequency dependent loss factor and material stiffness, implying that the piezocomposite becomes less stiff and changes into lossy material similar to viscoelastic in behavior. In cantilever beam damping application, the poling direction is generally along the thickness of the beam. Therefore, this is mainly transverse type of electromechanical coupling effect. Since the values of transverse electromechanical coupling coefficient are small, the variation of piezocomposite stiffness with frequency is negligible. Hence ignoring the frequency dependence of the piezocomposite stiffness would not significantly affect the Simulink results.

RESULTS AND DISCUSSION

Passive damping is achieved by energy dissipation through the resistor, which is not uniform but follows an "S" path as shown in Figure 4. This can be attributed to the RC type of electrical circuit. The vibratory motion of the beam generates a sinusoidal current in the piezoelectric material. Under the action of this varying current, the charge stored in the capacitor fluctuates in a cyclic manner. During the capacitor-charging period, the resistor draws less current leading to a lower energy loss, whereas during the capacitor-discharge period, the resistor draws the capacitor discharge current in addition to the piezoelectric current resulting in higher energy loss. During the cycle although there is a periodic variation, but over a range, the system energy displays a regular exponential type decaying behavior.

Figure 5 shows the beam tip amplitude. The damping ratio can be determined from the logarithmic decay of the amplitude and it is found to be same as the damping ratio calculated from the material loss factor [13].

The results of loss factor and system damping ratio due to piezocomposite device only are shown in Table 4. Variation of damping ratio with volume fraction is plotted in Figure 6. Damping ratio is observed to increase with rising volume fraction. This is an obvious relationship because as the piezoceramic fiber content goes up more energy will be available for dissipation in the resistor. At low volume fractions, the participation of piezoelectric fibers in the load bearing pattern is to a lesser extent and hence the damping ratio is low. On the contrary as the volume fraction rises, the involvement of piezoelectric fibers increases resulting in higher damping ratios. Since the inherent material damping in the aluminum beam and the structural damping at the fixed end are very small (0.0002), the additional damping provided by the piezoelectric tool goes up to a maximum of 0.42%. This being transverse loading configuration, higher damping ratios are not possible, on the other hand in structural systems where longitudinal arrangements are adopted higher damping ratios can be obtained.

When compared with the damping results from single degree of freedom system [4]. the damping ratios and its dependence on resistance display excellent agreement with each other. One source of error could be the assumption of frequency independent piezocomposite stiffness.

Table 4: Damping ratio by energy and single DOF method.

% Vol. Fract.	Loss Factor By Energy	Damping ratio By Energy	Damping ratio By Single DOF
15%	0.000485	0.000242	0.000209
25%	0.000532	0.000266	0.000244
30%	0.000546	0.000273	0.000261
40%	0.000600	0.000300	0.000294
50%	0.000710	0.000355	0.000362
60%	0.000903	0.000452	0.000480
100%	0.008420	0.004210	0.004570

CONCLUSIONS

The "ANSYS" and "SIMULINK" softwares are used to investigate the dynamic behavior of the piezoelectrically damped system. The computed damping ratio is seen to be in agreement with the single degree of freedom model results that were experimentally validated elsewhere [4]. To maximize the effectiveness of the piezocomposite damper, it has to be bonded at the location of high strain so as to extract maximum energy out of the system. For the cantilever beam it is the portion near the fixed end that gives the utmost damping. Vibrating cantilever beam proximity to transverse loading leads to low damping ratios, but in cases where longitudinal loading is possible much higher damping ratios can be achieved. It is observed that the higher volume fraction piezocomposites are considerably more effective dampers due to the relatively higher dominance of piezoceramic fibers than the lower volume fraction composites. Finally, in place of a cantilever beam, any type of vibratory structural system can be modeled by the ANSYS and SIMULINK combination to determine the damping introduced by the piezoelectric material.

REFERENCES

[1] Forward R. L., "Electronic Damping of Vibrations in Optical Structures" Journal of Applied Optics, 1979, pp 690-697.

[2] Crawley E. F. and Deluis J., " Use of Piezoelectric Actuators as Elements of Intelligent Structures" AIAA, 25, (10), pp1373-1385, 1987.

[3] Hagood N. W. and Crawley E. F., "Experimental Investigation of Passive Enhancement of Damping for Space Structures" Journal of Guidance, Control and Dynamics, Nov-Dec 1991, pp. 1100-1109.

[4] Hagood, N. W. and Von Flotow, A. H., "Damping of Structural Vibrations with Piezoelectric Materials and Passive Electrical Networks" Journal of Sound and Vibration, April-May, pp 243-267, 1991.

[5] Nasif A. D., Jones D. I. and Henderson J. P. "Vibration Damping" John Wiley, 1985.

[6] Website www.matweb.com

[7] "ANSYS", Version 5.6, SAS Inc., 2001.

[8] "IEEE Standards on Piezoelectricity" 176-1978.

[9] Nakhwa S., "Effective Properties and Vibration Damping of 1-3 Piezocomposites" M.S. Theses, Tufts University, 2002.

[10] Mason W. P. "Piezoelectric Crystals and Their Applications to Ultrasonics" D. Van Nostrand Company, 1950.

[11] "SIMULINK" Version 5.3, Mathworks Inc. 2000.

[12] Page L. and Adams N. I. "Principles of Electricity" D. Van Nostrand, 1958.

[13] De Silva C. W. "Vibration Fundamental and Practice" CRC Press, 2000.

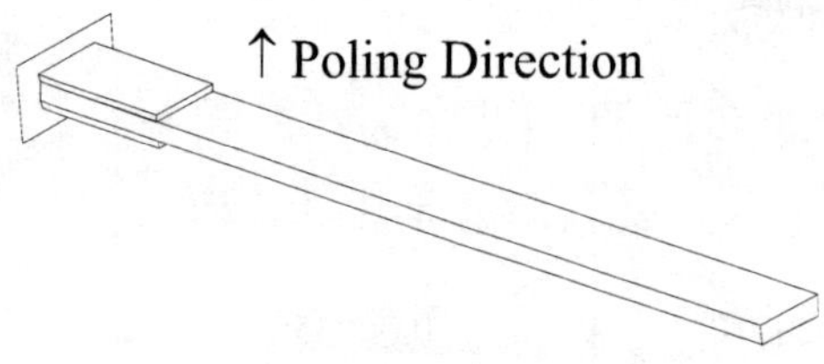

Figure 1: Cantilever beam structure with piezocomposite dampers.

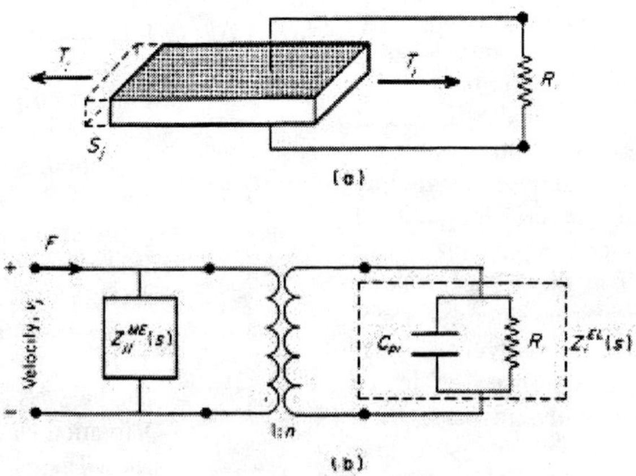

Figure 2: Uniaxially loaded (a) shunted piezoelectric model and (b) its network analog.

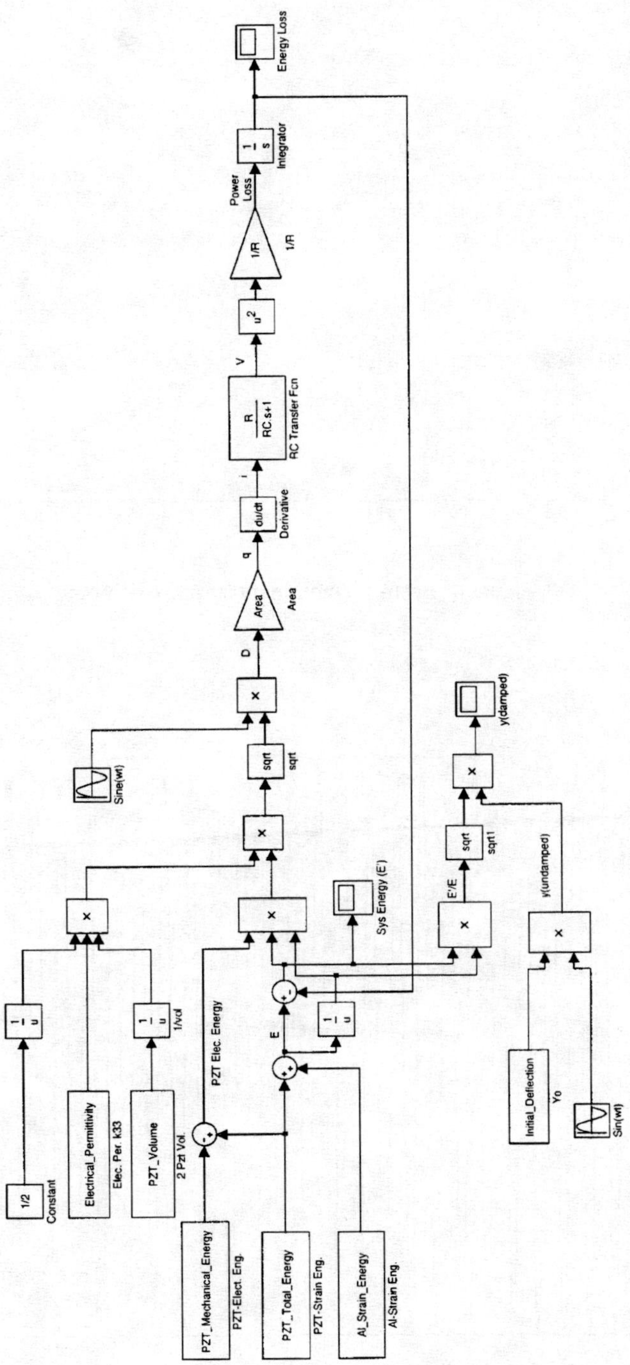

Figure 3: Block diagram of resistive piezoelectric shunting.

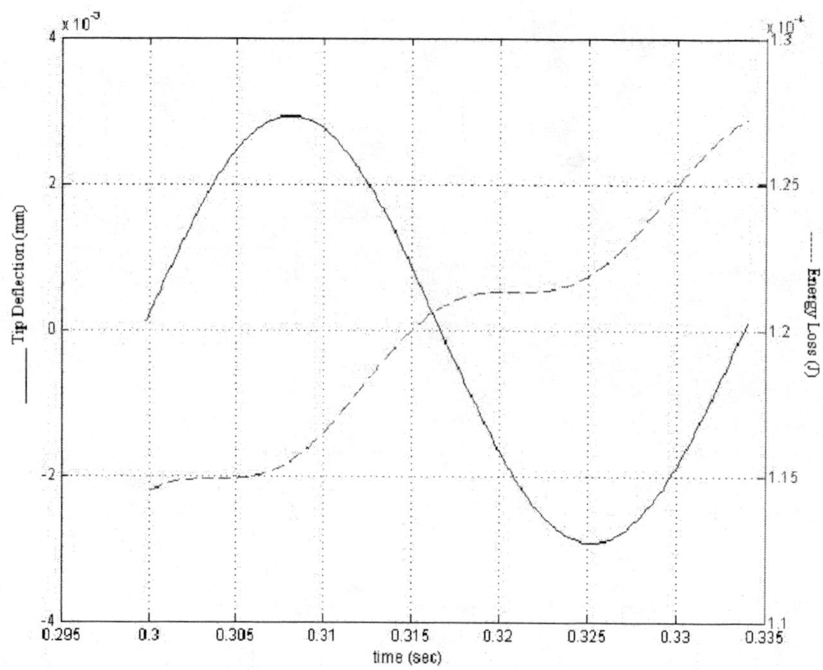

Figure 4: Beam tip motion under piezo damping.

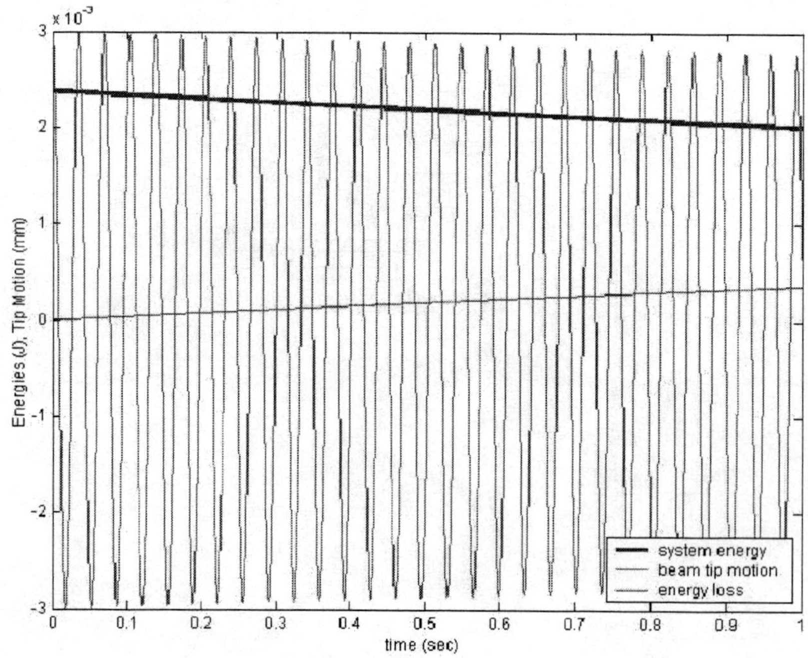

Figure 5: Resistor energy dissipation characteristics.

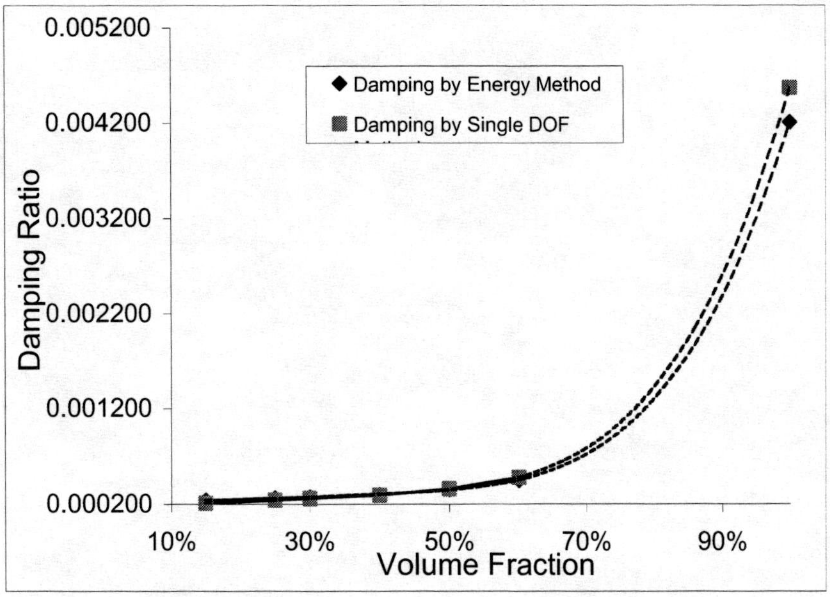

Figure 6: Dependence of damping ratio on volume fraction.

Proceedings of IMECE04
2004 ASME International Mechanical Engineering Congress and Exposition
November 13-20, 2004, Anaheim, California USA

IMECE2004-59572

Application of Operational Modal Analysis to a Machine Tool testing

Emmanuil Kushnir, Hardinge Inc., Elmira, NY

ABSTRACT

Modal analysis testing of a mechanical structure is performed usually by artificial excitation of a structure and measuring input forces and output responses of a mechanical system. The excitation is either transient (impact hammer testing), random, burst-random or sinusoidal (shaker testing). The modern signal processing tools enable to determine properties of a mechanical structure such as resonance frequencies, damping ratios, and mode patterns by measuring the response of the structure without using an artificial excitation. The advantage of this technique is that modal parameters of a structure may be evaluated while the structure is under actual operating conditions. That will allow developing a model within true boundary conditions and actual force and vibration levels.

The machine tool structure characteristics that effect productivity and quality have to be evaluated by testing. These characteristics include natural frequencies, modes of vibration, and external sources of high level vibration. Not all modes of machine tool structure effect machine quality. As a result only the modes that are excited during cutting have to be taken in the account. This approach narrowed the frequency range, which has to be considered in test.

The machine tool during cutting and/or idling is loaded by a set of external and internal exciting forces. Spectrum, frequency range and application points of these forces are unknown in many cases. Under these exciting forces the vibration between the tool and workpiece, and vibration of machine tool components are sums of many independent vibrations and may be considered as stationary random processes. This assumption allows applying the theory of stationary random processes to machine tool dynamic testing during cutting.

Several characteristics of random processes are used to separate harmonic vibration from narrow-band random vibration at natural frequencies. The spectral analysis of machined surface profiles and its correlation with observed vibration allows choosing modes that have to be developed. The analysis of these modes provides a basis for machine tool structure improvement. The proposed experimental approach was verified by experiments at different machine tools. Results of these tests are presented in the paper.

Keywords: machine tool, operational modal analysis, vibration, cutting, natural frequency, stationary random processes.

TESTING CONDITIONS

The general flow of data in machine tool structure operational analysis is presented in Figure 1. Some of the steps and measured parameters are optional. The analysis of input data was performed by standard procedures, used in spectral analysis (Bendat, 1971).

The testing condition is defined by the goal of an experiment. In the case when machine tool does not cut part with required efficiency and quality the test has to be perform during these specific cutting conditions. In case of general dynamic testing the goal is to obtain data for future development, machine modeling, etc. In this type of testing, most common for particullar machine cutting conditions have to be chosen.

The aim of testing is to define all modes of vibration and ambient sources of vibration that contribute to quality of machining.

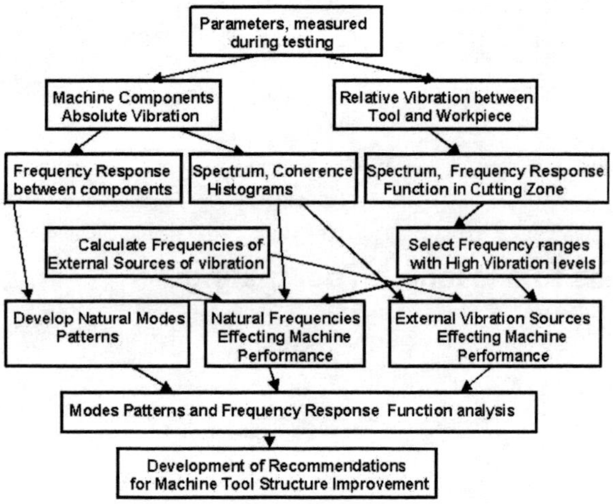

Fig.1. Data flow and analysis procedures, performed during operational dynamic testing of machine tools.

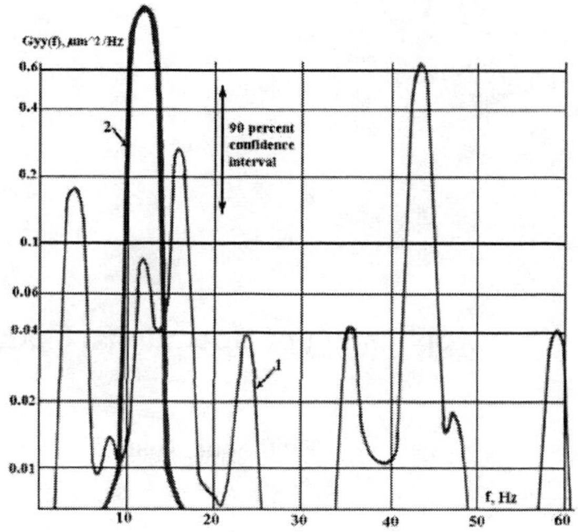

Fig.2. Gear-grinding machine tool spectrum of vibration in cutting zone: 1-230 double stroke/min; 2 –110 double stroke/min.

It is important to keep in mind that not all modes of machine tool structure in an analyzed frequency range effect machine quality. Only the modes, which are excited during cutting, have to be taken in the account. This approach narrowed the frequency range, that has to be considered, and defined the vibration characteristic that are used in proposed experiment technique.

MEASUREMENT PROCEDURE

At the first step, the relative vibration between tool and workpiece have to be measured. This data provides information, required to define frequency range for modal testing and, in general, to define frequencies of high level vibration in cutting zone that may affect quality of machining.

As an example at Figure 2, curve 1, is presented a spectrum of vibration in the cutting zone of the vertical gear-grinding machine tool.

The cutting condition of the test are as follows: number of strokes – 230 double stroke/min; length of the stroke – 90mm; number of gear teeth – 50, gear width – 40 mm, pitch – 6mm, grinding wheel speed – 2070 RPM. The high level of vibration has place in low frequency range at 4, 8, 12, 16 Hz . Other frequency ranges of interest are 23, 35, 43, and 58 Hz.

The known frequencies of external sources of vibration are: slide double stroke frequency – 4 Hz (230 double stroke/min), grinding wheel unbalance frequency – 35 Hz. Because of slide reverse motion the harmonics of double stroke frequency 8,12, 16 Hz and higher are presented in spectrum and may be mixed with other external sources of vibration and structure natural frequencies.

NATURAL FREQUENCIES and EXTERNAL SOURCES of VIBRATION

The next important step is to distinguish vibration at natural frequencies and frequencies of external sources of vibration and its harmonics.

From theoretical point of view no formal technique exists to distinguish narrow-band random data and signal that is a sum of a periodic component and random noise. From practical point, it may be done if the theoretical analysis is supported by physical evidence. In the proposed approach, different theoretical characteristics of random processes are used to check out type of observed vibration, and the frequencies of exciting external vibration are compared with analysis. If this information is not enough to make a defined conclusion a test has to be performed under different cutting conditions.

Several characteristics of random processes may be used to separate harmonic vibration from narrow-band random vibration.

The coherence function $\gamma^2(f)$ shows the correlation between vibrations in different points of the machine. At natural frequencies the coherence function has to be close to 1 in relative wide range because the frequency response function in this range has a width proportional to damping. In this case the vibration near natural frequency will have relative high level in wide range. In situations when vibration is developed by single source of periodic vibration the high level of vibration may be observed in very narrow range. The coherence function between vibration of the column and workpiece for the gear-grinding machine in consideration is presented in Fig. 3. The width of area, where coherence function is close to 1 is a lot wider near frequency 43 Hz, than near frequency 35 Hz. It means that frequency of 43 Hz may be natural frequency of machine structure. The frequency of 35 Hz is a frequency of external source of vibration, which in this case is an unbalance of grinding wheel, rotating with 2070 RPM.

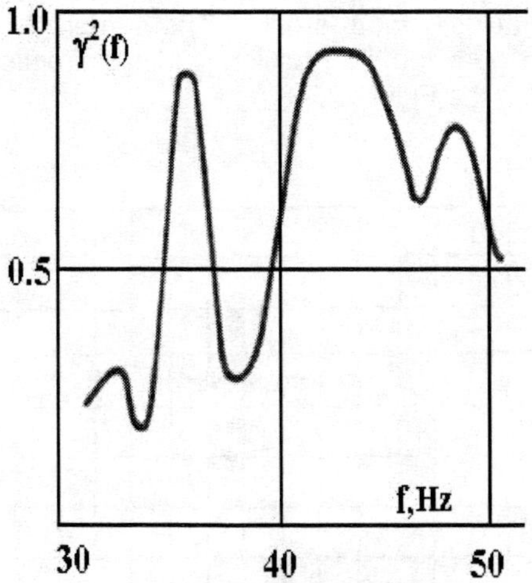

Fig. 3. The coherence function between vibration of workpiece and column of a gear-grinding machine tool.

The other way to support this conclusion is to compare a histogram of vibration near these frequencies after narrow band filtration. The vibration at frequency 35 Hz will have a histogram with the shape, which is close to histogram of sinus wave (see Fig.4a). On the other side, the histogram of vibration near 43 Hz has shape close to one of narrow random noise (see Fig. 4b).

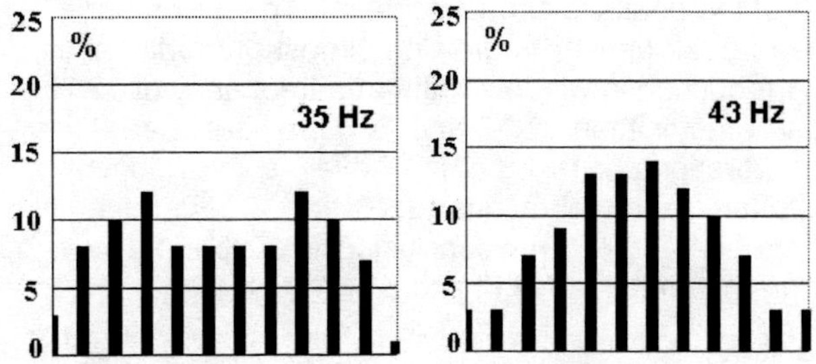

Fig. 4. Histogram plots gear grinding machine vibration at different frequencies: (a) - 35 Hz, (b) - 43 Hz.

It is known (Jenkins, Watts, 1968) that estimation of spectrum density for any stationary random process will be changed with change of spectral window width. More important, as it follows from simple physical analysis the change in density will be to some extent different for random process and for a sum of a harmonic with random noise. Two parameters will be changing different ways: width of spectrum peak and maximum value of the peak.

Let's define the width dF of spectrum peak as a distance in Hz between points, where level of energy in spectrum around peak equal 0.5 of peak level (Bendat, 1971). For a process with a periodic component the width of spectrum peak dF will be equal to the width of spectral window. In the case of narrow random vibration the width of spectral window practically does not effect width of spectrum peak. It will be the same as long the spectral window will be wider than width of frequency response function peak. Width of frequency response function $\mathbf{dW = f_0{*}\lambda{*}\pi}$, where λ - logarithmic decrement, f_0 – natural frequency.

The energy of vibration for a sinusoidal component is concentrated at one frequency. Thus means that with decrees of spectral window the maximum estimation of density at frequency of interest has to grow, to provide the same energy inside the window. In the case of random vibration the energy is distributed in the range of frequency response function peak, and with decrease of the width of the window the maximum estimation of density at frequency of interest has to grow slowly in comparison with a former case.

The example of application of this methodology is presented in Figure 5. The two-column vertical turning center was tested during cutting to obtain experimental data.

The vibration of column in the range 14 – 30 Hz was analyzed with different spectral windows Be: 9.5, 2.9, and 1 Hz. The width of the peak is changing near frequency 22 Hz and is close to the bandwidth of spectral window in cases of Be =1, and Be = 2.9 Hz. Near frequencies of 16 and 26 Hz the width of the peaks is almost constant. The maximum value of peak density is growing almost proportional to the reduction in window width at a frequency of 22 Hz, and faster than at 16 and 26 Hz. This means that vibrations at 22 Hz have a harmonic component. The vibrations at 16 and 26 Hz have random character. This conclusion is supported by histograms presented in Figure 6.

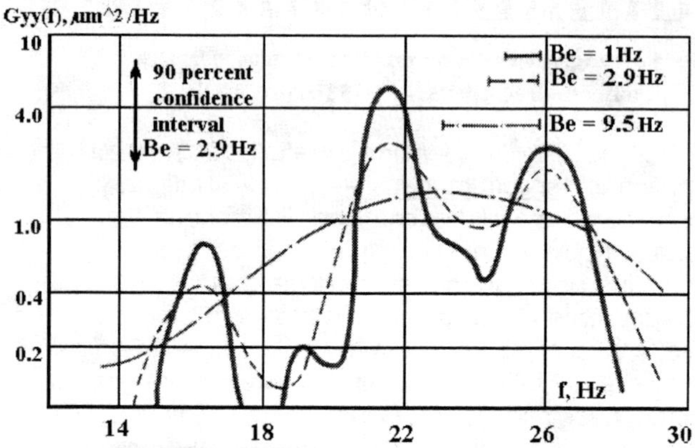

Fig. 5. Spectrim of vibraton in cutting zone of a turning center computed with different spectral window Be.

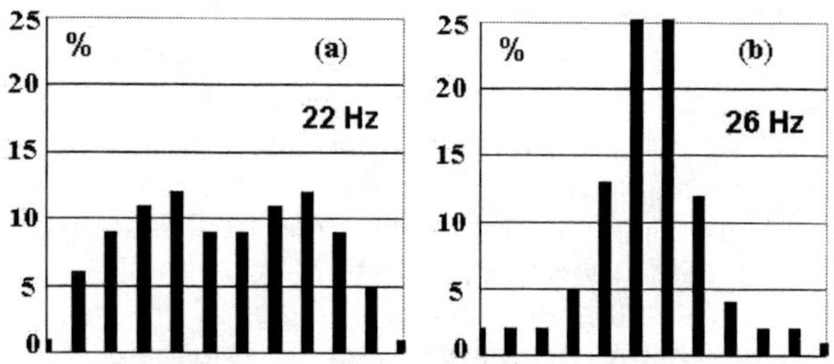

Fig. 6. Histogram plots of a turning center vibration at different frequencies: (a) - 22 Hz, (b)- 26 Hz.

The comparison of electric motors speed with presented data showed that vibrations at 22 Hz are stipulated by an unbalance of the main drive motor that was running at 1350 RPM. The frequencies 16 and 26 Hz are natural frequencies of machine tool structure. This conclusion was also supported by smooth shape of coherence function in the range of 16 and 26 Hz.

In the case when a periodic exciting force coincidences with the natural frequency of a mechanical structure, the described methods do not work if the sinusoidal component is not fairly intensive relative to random portion of vibration. It is justified in this case to change the cutting condition and compare the spectrums of vibration. For example, curve 2, at Figure 2, represents vibration of the gear-grinding machine under 110 double strokes per minute. In the observed frequency range one peak at 12 Hz is substantially bigger than any other ones. It is obvious, that in this case, one of the excited force harmonics coincides with the natural frequency of the structure near 12 Hz.

It makes sense to mention that the stable width of the spectrum peak at a natural frequency, obtained during describe procedure of changing spectral window width, may be used for evaluation of the mechanical system damping at this particular mode.

MODE PATTERNS

The next step in operational dynamic analysis of machine tool structure is mode patterns development. One of the advantages of operational dynamic analysis is the fact that not all the modes of a machine tool structure have to be analyzed. The patterns belonging only to these modes that are excited during the cutting tests and, in first place, which have direct impact at machine tool machining performance have to be developed and analyzed. That means that only characteristics that represent machine tool cutting performance have to be taken in consideration at this step of testing and analysis.

In a vertical gear grinding machine vibration between the grinding wheel and workpiece has direct effect at gear tooth profile. The sample of a profile of the gear tooth in the longitudinal direction is presented in the Figure 7.

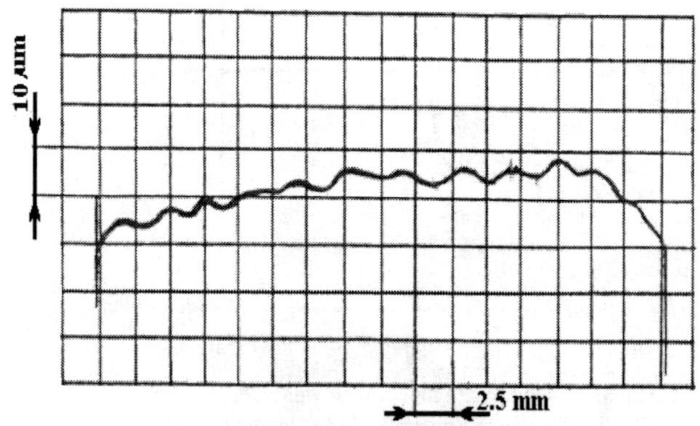

Fig. 7. Profile of a tooth, machined at gear grinding machine. Tooth length =40 mm, wheel speed 2070 RPM, 220 stroke/min

The gear was ground under conditions described earlier. Simple calculations showed that frequency of the wave of curve, presented at Figure 7 equals 90mm*230/(60*2)/4mm ≈ 43 Hz. Thus the natural mode of machine tool vibration at 43Hz has direct impact at the quality of machining and has to be analyzed and improved first.

The simplified pattern of machine vibration at this mode was obtained by using only 12 points for vibration measurement. The point at the column with high amplitude was chosen as a reference point (M. Batel, 2002, Bendat J.S., Piersol A.G., 1971). The frequency response functions (FRF) between this point and other points were computed. The modes of vibration were estimated from spectral densities calculated with the assumption of white noise input and lightly damped structure (Kaminskayia V.V., 1967, Kushnir E. and others, 2001). The developed mode of vibration of the vertical gear-grinding machine at natural frequency 43 Hz is presented in the Figure 8.

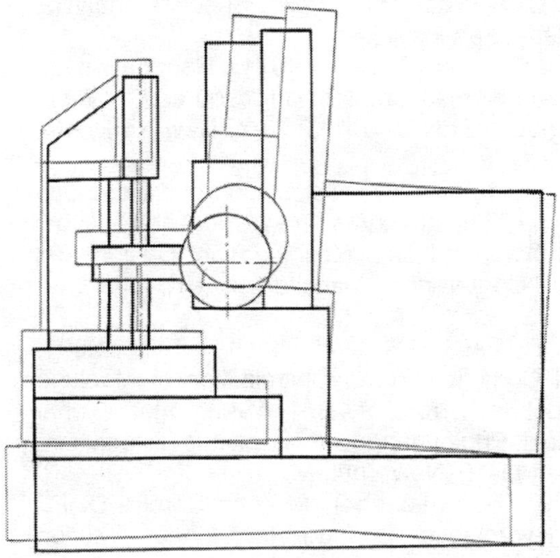

Fig.8. Gear-grinding machine mode of vibration at frequency 43 Hz.

It is may be observed from Figure 8 that the mode of vibration at 43 Hz is stipulated by base bending. The mode pattern is determined by a heavy column at one side of the base and relative light workpiece mounting support at the other side.

The recommendation to increase the base stiffness and stiffness of the joint between column and base allowed for the increase of static stiffness and natural frequency of this particular mode, and decrease level of vibration in the cutting zone. These developments increased quality of gear tooth grinding.

The other case shows effects of machine component characteristics regarding machine performance. The frequency response function of the lathe with linear recirculating ball bearings is presented in Figure 9, curve 1. This machine has the highest dynamic compliance at a natural frequency of 160 Hz.

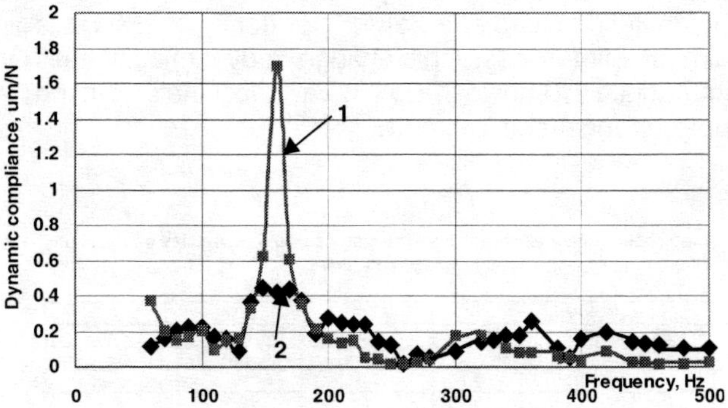

Fig.9. Frequency response function of a lathe. Curve1- machine with linear ball guides, curve 2-with hydrostatic guides.

Brass samples were cut at this lathe at 3 different RPM (revolution per minute). Roundness values and roundness harmonics, obtained from measurement of these brass samples, are presented in Table 1.

Table 1

Correlation between roundness harmonics and frequencies of vibration in the lathe with linear ball guides

RPM			RPM		
1200	1600	2400	1200	1600	2400
Cutting Roundness Harmonics			Frequency, Hz	Frequency, Hz	Frequency, Hz
2	2	2			
3			60		
		4			160
5			100		
	6			160	
	7			187	
8		8	160		320
9			180		
10			200		
ROUNDNESS,um					
0.46	0.53	0.65			

The frequency of 160 Hz showed up in cuts at all 3 RPM. That means that the dynamic compliance of the machine at 160 Hz has direct impact at the roundness of the part, machined at this lathe. The finite element analysis of the lathe structural model and testing showed that the vibration at 160 Hz during cutting is defined by a complex mode of vibration, involving both X and Z axis slide.

The cutting quality of the machine was improved by substitution of linear recirculating ball bearings with hydrostatic guides (Wasson, K. L. and others, 2002). These hydrostatic guides have similar to linear ball bearings static stiffness and a lot higher damping.

The FRF of the lathe with hydrostatic guides is presented in Figure 9, curve 2. As it follows from Figure 9 the dynamic compliance of the machine with hydrostatic guides at a natural frequency of 160 Hz is almost 4 times less. This change in dynamic stiffness immediately showed up in a machine's cutting performance (see Table 2).

Table 2

Correlation between roundness harmonics and frequencies of vibration in the lathe with hydrostatic guides

RPM			RPM		
1200	1600	2400	1200	1600	2400
Cutting Roundness Harmonics			Freqiency, Hz	Freqiency, Hz	Freqiency, Hz
2	2	2			
3	3	3	60.0	80.0	120.0
	4			106.7	
		5			200.0
9			180.0		
ROUNDNESS,um					
0.08	0.25	0.14			

The data, presented in Table 2, was obtained from brass samples cut at 3 different RPM's under the same condition as that of samples presented in Table1. This data does not show any correlation between roundness harmonics and vibration for the machine with hydrostatic guides at frequency 160 Hz.

CONCLUSION

The presented examples of experimental analysis of machine tools structures under operational load conditions show that purposeful analysis of machine vibration allows the obtainment of practical application results with minimum expenses of the time and resources.

Dynamic testing of a machine tool during cutting allows to perform troubleshooting in real operational conditions directly on the shop floor. Only accelerometers are used in testing. This allows, in many cases, performed testing without interruption of the manufacturing process.

REFERENCES

M. Batel, 2002. Operational Modal Analysis – another Way of Doing Modal Testing. Sound and Vibration, August 2002, pp. 22-27.

Jenkins,G.M, Watts D.G., 1968. Spectral Analysis. Holden-Day, San Francisco.

Bendat J.S., Piersol A.G., 1971, Random data. Analysis and measurement procedures. Wiley-Interscience. ISBN 0-471-06470-X, New York, 407 pp.

Kaminskayia V.V., 1967, "Calculation and Research on Machine Tool structures and foundations," Proc. of the 8-th International Machine Tool Design and Research Conference, Manchester, UK, pp. 139-161.

Kushnir, E., Patel M.R., and Sheehan T. M., 2001, "Material Consideration in Optimization of Machine Tool structure," Proc. of 2001 ASME International Mechanical Engineering Congress and Exposition, November 11-16, New York, NY.

Wasson, K. L., Sheehan T. M., and Soroka D. P., 2002, "Hydrostatic Bearing for Linear Motion Guidance", patent pending.

Proceedings of IMECE04
2004 ASME International Mechanical Engineering Congress and Exposition
November 13-20, 2004, Anaheim, California USA

IMECE2004-59573

Determination of machine tool frequency response function during cutting.

Emmanuil Kushnir, Hardinge Inc., Elmira, NY

ABSTRACT

The dynamic compliance (frequency response function - FRF) of a machine tool structure in the cutting zone under a cutting load is one of the major dynamic characteristics that define a machine's cutting performance. The roundness and surface finish define the quality of the manufactured parts. These characteristics are developed during finishing and semi-finishing cuts. The knowledge of machine tool dynamic compliance, defined in these steady-state cutting conditions, ensures parts quality and increase in machine tool productivity.

The dynamic compliance is usually evaluated in tests, which are performed by means of hammers or vibrators (exciters). During these tests the machine does not cut and the machine components do not move relative to each other. The loads in the machine during cutting are defined by different internal and external sources that are acting in different points of the machine and in different directions. The real spectrum and frequency range of these forces is unknown. Experimental data acquired by different types of tests clearly show the difference in dynamic compliance for the same machine tool during cutting and idling. The machine tool dynamic tests performed by different types of external exciting devices do not take in consideration the real load conditions and interactions of moving components, including the cutting process itself and external sources of vibration.

The existing methods of experimental evaluation of machine tool dynamic compliance during steady-state cutting condition require dynamometers to measure the cutting force and a special sensor to measure relative displacement between the cutting tool and workpiece. The FRF that is computed from these measurements represents a dynamic characteristic of the close loop system (machine structure and cutting process) and only under certain conditions may be considered as FRF of machine tool structure itself.

The theory of stationary random processes allows defining the cutting conditions, under which the obtained data represent the FRF of machine tool structure, and provide estimations of random and bias errors of this evaluation. The simplified methodology of FRF estimation, based only on measurement of the spindle and tool vibration, is also presented in this paper. This methodology is used on an assembly line to obtain FRF for machine tools performance comparison and quality assurance.

Keywords: machine tool, dynamic compliance, frequency response function, cutting process, vibration, stationary random processes.

NOMENCLATURE

$W_c(f)$ –frequency response function (FRF) of cutting process;
$W_{es}(f)$ – FRF of machine tool structure;
W_φ (f) –FRF of machine tool structure under external disturbance;
$K_{es}(f)$ - dynamic stiffness of the machine tool structure;
f - frequency, Hz;
p(t - dynamic component of cutting force;
y(t)- relative vibration between tool and workpiece during cutting and idle;
$\varphi(t)$ - external disturbance during cutting and idle;

z (t) - disturbance, acting on working process;
$G_{lm}(f)$ - power spectral (or cross-spectral) density function (power spectrum of process);
l,m = (p,y, φ) – input and output components;
Eb - normalized bias error;
w - width of the slots;
a - height of the slots;
τ_0 - width of slots in the time domain;
T - distance between the slots in the time domain;
K - ratio τ_0 / T.

INTRODUCTION.

The dynamic stiffness (compliance) of machine tool structure in cutting zone under cutting load is one of the major characteristics of a machine that define its cutting performance. Under cutting performance in this case we will mean:

- relative level of vibration in cutting zone during finish cut, which define the roundness and surface finish of the part;
- maximum depth of cut without chatter.

Usually the chatter are defined by one or two modes of vibration, but the roundness and surface finish are defined by vibration in a wide frequency range, which include the compliance of the machine tool at natural frequencies and frequencies of external exciting forces. This means that the chatter test, which is often used to provide experimental evaluation of machine tool performance, does not represent dynamic stiffness of a machine in the required frequency range. The chatter test is a simple and a quick test that may be performed on assembly line or in the shop without any special measuring tools. The results of these chatter tests may be compared on the relative basis as a maximum depth of cut under certain cutting condition or recalculated to a dynamic stiffness at mode of vibration at which chatter occur *occurs* (Kaminskaya, Kushnir, 1979).

In present time the experimental dynamic stiffness is usually evaluated in tests, those are performed by means of hammer or vibrators (exciters). During these tests a machine does not cut and make things worse, usually the machine components do not move relative to each other. The loads in the machine during cutting are defined by different internal and external sources that are acting in different points of the machine and different direction. The real spectrum and frequency range of these forces is unknown. Meaning, that machine tool dynamic tests performed by different type of external exciting devices do not take in consideration real load conditions and interaction of moving components, including the cutting process itself and external sources of vibration (Tlusty, Moriwaki, 1976).

The evaluation of machine tool dynamic compliance during cutting require dynamometers or special tool to measure the cutting force and sensors to measure relative displacement between tool and workpiece. The dynamic compliance, computed from this data, represents a dynamic characteristic of the close loop system (machine structure and cutting process) and only under certain conditions may be considered as FRF of machine tool structure itself (see below). In the case when dynamometer is used the obtained FRF represents mechanical system, combined from machine tool structure and dynamometer.

The vibration of machine tool structure is developed by simultaneous action of a large number of disturbances, applied in different point of the machine. These disturbances have different nature (cutting forces, motors, spindle and fans unbalance, hydraulic pumps, etc.). The vibration of machine tool components under these circumstances may be considered as stationary random processes.

THE GENERAL THEORY.

The dynamic system of the machine tool may be presented by a well-known single closed loop system (see Figure 1).

The machine tool structure is presented in this loop

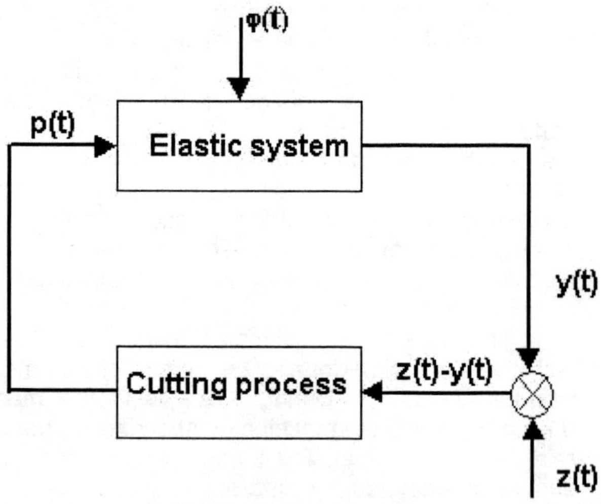

Fig1. Close loop representation of machine tool structure

by an elastic system (ES). A cutting process (CP) represents the feedback in this loop. In cutting, this system is loaded by two independent disturbances. The external disturbance φ(t) is acting on the machine tool structure. All variation in cutting process conditions, such as the variable depth of a cut and/or hardness of a workpiece, may be presented by disturbance acting on the cutting process z (t). Here t – is current time.

In accordance with the theory of stationary random processes [Bendat, 1971) the following equations will represent the system motion:

$$G_{py}(f) = W_{es}(f)*G_{pp}(f) \; + W_{\varphi}(f)*G_{p\varphi}(f) \qquad (1)$$

$$G_{\varphi p}(f) = W_c(f)*G_{\varphi y}(f) \; + W_c(f)*G_{\varphi z}(f) \qquad (2)$$

$$W_{\varphi}(f)*G_{\varphi\varphi}(f) = [1 \; + W_c(f)*W_{es}(f)]*G_{\varphi y}(f) \qquad (3)$$

In the case when the machine is running, but cutting is not performed, the system's motion my be represent by the following equation

$$G_{yy}\,\text{idle}\,(f) = |W_{\varphi}(f)|^2 *G_{\varphi\varphi}(f) \qquad (4)$$

During the test one may measure:

- dynamic component of cutting force p(t);
- relative vibration between tool and workpiece y(t) during cutting and idle.

The application points of external disturbance $\varphi(t)$ are unknown and in correspondence the FRF $W_{\varphi}(f)$ cannot be found.

It is appropriate to assume that that the external disturbance $\varphi(t)$ and FRF $W_{\varphi}(f)$ during cutting are the same as in idle. The external disturbances $\varphi(t)$ and z (t) are independent and as a result the function $G_{\varphi z}(f) \equiv 0$. Based on this assumption and considering $G_{\varphi p}(f) = G^{*}_{p\varphi}(f)$ (the superscript * indicates a complex conjugate function) the following expression may be obtain from equations (1)-(4)

$$W_{es}(f) = G_{py}(f)/G_{pp}(f) ++\{W_c(f)/[1+ W_c(f)*W_{es}(f)]\}^* \{ G_{yy}\,\text{idle}\,(f)/G_{pp}(f)\} \qquad (5)$$

The FRF $W_{es}(f)$, defined by equation (5), represents both the modulus and the phase of the complex function $W_{es}(f)$ and, for example, is necessary for chatter condition evaluation.

It follows from equation (5) that in order to evaluate FRF $W_{es}(f)$ from a experimental data, which included force and relative vibration measurements, one has to know FRF of the cutting process $W_c(f)$. It is the reason that only the first member in equation (5) is used in practice for $W_{es}(f)$ experimental evaluation (Moriwaki, Ivata,1976)

$$W_{es}(f) = G_{py}(f)/G_{pp}(f) \qquad (6)$$

In this case the second member in equation (5) may be considered in accordance with the theory of stationary random processes as component of systematic (bias) portion of the error of the FRF estimation. As it follows from (5) this error will decrease as the depth of a cut increases and disturbance, acting on working process z(t), increases. It may be also shown that the random error of the FRF evaluation (interval of confidence) will be reduced if the disturbance, acting on working process z(t), increases.

<u>Example.</u>

The FRF of a lathe was evaluated during cutting of a steel part in chuck. Diameter of the part was 100 mm, length 70 mm. Cutting conditions: speed 100 m/min, feed rate 0.15 mm/rev. Cutting was performed

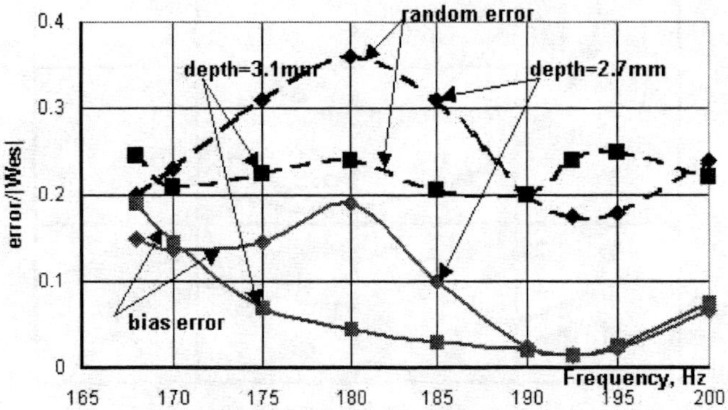

Fig. 2. Error of machine tool FRF experimental evaluation with different depth of cut.

with a tool, which has a shank with force gages in it. The test was performed with two depths of cut 2.7 and 3.1 mm. The errors of evaluation of FRF under these conditions are presented in Figure 2.

As it may be seen from the Figure 2, in region of frequency 180 Hz the increase in depth of cut permits, in full agreement with theory, reduce the bias error and narrow the interval of confidence. The plot of actual FRF obtained in test is presented in Figure 3, for the case of cutting with the depth 3.1 mm. The maximum depth of the cut, obtained in chatter test, was close to one obtained from calculation performed by using experimental chart, presented in Figure 3.

THE SIMPLIFIED APPLICATION APPROACH

The FRF $W_{es}(f)$ defined by equation (5) represents both the modulus and the phase of the complex function $W_{es}(f)$ and is necessary , for example, for chatter condition evaluation. In many other cases only the absolute value $|W_{es}(f)|$ or gain of FRF $W_{es}(f)$ may be used for machine tool dynamic stiffness evaluation. In addition to assessment of this characteristic that follows from equation (5), the other formula may be obtained from spectral density $G_{yy}(f)$ representation, based on closed loop system motion :

$$|W_{es}(f)|^2 = G_{yy}(f)/G_{pp}(f) - \{[1 - |W_c(f)*W_{es}(f)|^2]/ |1 + W_c(f)*W_{es}(f)|^2\}^* \{ G_{yy}\,\text{idle}\,(f)/G_{pp}(f)\} \qquad (7)$$

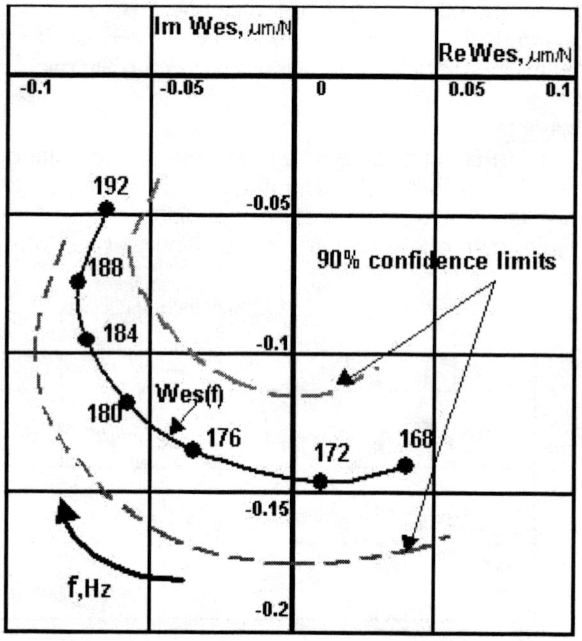

Fig.3. Frequency responce function of lathe, defined during cutting with depth 3.1 mm.

As in the previsions case only first member in equation (7) is used in practice for $|W_{es}(f)|$ experimental evaluation

$$|W_{es}(f)|^2 = G_{yy}(f)/G_{pp}(f) \qquad (8)$$

The second member in equation (7) may be considered in accordance with the theory of stationary random processes as component of systematic (bias) portion of the error of the FRF estimation by equation (8).

As a further convenience, the error will be estimate in the terms of fractional portion of frequency response function. This is done by dividing the error by the module of the FRF to obtain a normalized error Eb

$$Eb=\sqrt{G_{yy}(f)/G_{pp}(f) - \{[1-|W_c(f)*W_{es}(f)|^2]/|1+W_c(f)*W_{es}(f)|^2\}^* \{G_{yy}\ idle\ (f)/G_{pp}(f)\}/\{G_{yy}(f)/G_{pp}(f)\}} - 1 =$$
$$\sqrt{1-\{[1-|W_c(f)*W_{es}(f)|^2]/|1+W_c(f)*W_{es}(f)|^2\}^* \{G_{yy}\ idle\ (f)/G_{yy}(f)\}} - 1 \qquad (9)$$

During steady state finishing cuts, the depth of a cut is relatively small compared to the depth of a cut at which chatter occurs. It means that module

$$|Wc(f)*\ Wes(f)| \ll 1 \qquad (10)$$

Under these conditions equation (9) my be rewritten as

$$Eb \ll \sqrt{\{1/(1-G_{yy}\ idle\ (f)/G_{yy}(f)\}} - 1 \qquad (9')$$

Under assumption, that level of relative vibration during cutting is a lot higher that the idle one

$$G_{yy}\ idle\ (f)/G_{yy}(f) \ll 1\ , \qquad (10')$$

the bias error may be estimated by inequality

$$Eb \ll 0.5\ G_{yy}\ idle\ (f)/G_{yy}(f) \qquad (9'')$$

The expression (8) requires measurements of the cutting force and relative displacement for $|W_{es}(f)|$ experimental evaluation. This is still not practical in many cases on the shop floor. From close loop representation of machine tool dynamic system the following two expressions may be wrote down also:

$$G_{pp}(f) = [|W_c(f)|^2]^* G_{z-y,z-y}(f) = [|W_c(f)|^2]^*\{G_{zz}(f)+G_{yy}(f)-[G_{zy}(f)+G_{yz}(f)]\} = [|W_c(f)|^2]^*\{G_{zz}(f)+G_{yy}(f)-2ReG_{yz}(f)\} \qquad (11)$$

$$G_{yz}(f) = \{[W_c(f)*\ W_{es}(f)]/[1+W_c(f)*W_{es}(f)]\}^*\ G_{zz}(f)\} \qquad (12)$$

During finishing cuts $|Wc(f)*\ Wes(f)| \ll 1$ and, as it follows from equation (12)

$$|G_{yz}(f)| \approx |W_c(f)*\ W_{es}(f)|\ *\ G_{zz}(f)$$

and

$$ReG_{yz}(f) \le |G_{yz}(f)| \approx |W_c(f)*\ W_{es}(f)|\ *\ G_{zz}(f) \qquad (12')$$

As long as this condition is satisfied, equation (11) may be used in practical application in the simplified form:

$$G_{pp}(f) = [|W_c(f)|^2]^* G_{z-y,z-y}(f) \approx [|W_c(f)|^2]^*\{G_{zz}(f)+G_{yy}(f)-2|W_c(f)*\ W_{es}(f)|\ *\ G_{zz}(f)\} \approx [|W_c(f)|^2]^*\{G_{zz}(f)+G_{yy}(f)\} \qquad (13)$$

Substituting $G_{pp}(f)$ in equation (8) by expression (13)

$$|W_{es}(f)|^2=G_{yy}(f)/G_{pp}(f)=G_{yy}(f)/\{[|W_c(f)|^2]^*[G_{zz}(f)+G_{yy}(f)]\} \qquad (14)$$

The main term of bias error of $|W_{es}(f)|$ estimated by Equation (14) may be found as a difference in estimations performed by Equation (8) and (14). As in the earlier case the error will be estimate in the terms of fractional portion of frequency response function. For a Wes $\neq 0$ and $|Wc(f)^* Wes(f)| << 1$, the normalized bias error is given by,

$$Eb = \sqrt{\{ G_{yy}(f) / \{[|W_c(f)|^2]^*[G_{zz}(f)+G_{yy}(f) - 2ReG_{yz}(f)]\} / \{G_{yy}(f) / \{[|W_c(f)|^2]^*[G_{zz}(f)+G_{yy}(f)]\}} - 1 =$$

$$\sqrt{[G_{zz}(f)+G_{yy}(f) - 2ReG_{yz}(f)]) / [G_{zz}(f)+G_{yy}(f)]} - 1 \leq$$

$$\sqrt{[G_{zz}(f)+G_{yy}(f) - 2|W_c(f)^* W_{es}(f)|^* G_{zz}(f)]) / [G_{zz}(f)+G_{yy}(f)]} - 1$$

$$\leq \sqrt{1 - 2|W_c(f)^* W_{es}(f)|} - 1 \leq |W_c(f)^* W_{es}(f)| \qquad (15)$$

The sum of bias errors, estimated by (9") and (15) under the assumptions that G_{yy} idle $(f)/G_{yy}(f) << 1$ and $|Wc(f)^* Wes(f)| << 1$, provide the bias error that will be obtained if the approximate equation (14) will be used for evaluation of $|W_{es}(f)|$ instead of original equation (7).

The variance (random) term of error of the estimate values may be evaluated by known approaches. In a real cutting test the coherence function between disturbance z(t) and y(t) is very close to 1 if the condition (10') is satisfied . As it is known (Bendat, 1971), in this case the random error will be small.

The reciprocal of dynamic compliance is defined as dynamic stiffness $K_{es}(f)$ of the machine tool structure. The $K_{es}(f)$ may be estimate from equation (12) as:

$$| K_{es}(f)| = |W_c(f)|^* \sqrt{1 + G_{zz}(f)/G_{yy}(f)} \qquad (16)$$

The expression (16) may be used for dynamic stiffness $K_{es}(f)$ experimental estimation if all conditions mention above are satisfied and a gain of the cutting process FRF is known.

In many cases there is a requirement to compare dynamic stiffness of similar machines, for example, as a part of quality assurance process at an assembly line. In this case the cutting process may be standardized for the machine tool dynamic stiffness evaluation. This means that insert, material of workpiece, feed rate, coolant, and etc. have to be the same. In this case the cutting process may be excluded from consideration and relative dynamic stiffness may be estimated by the equation

$$|K_{res}(f)| = | K_{es}(f)| / |W_c(f)| = \sqrt{1 + G_{zz}(f)/G_{yy}(f)} \qquad (17)$$

This means that the module of relative dynamic stiffness $|K_{res}(f)|$ may be defined experimentally by a known load (disturbance acting on cutting process) $G_{zz}(f)$ and measured relative displacement between tool and workpiece $G_{yy}(f)$. The value of $|K_{res}(f)|$ may be used for a machine tools dynamic stiffness comparison.

As it is known, disturbance acting on the cutting process may be part of a process, as in milling, or developed artificially, for example, in turning.

The disturbance, developed during finish turning process is not powerful enough to satisfy the condition (10'). The power of disturbance has to be increased. In turning, a disturbance may be developed by variation in the depth of a cut, originated from slots in workpiece. Let's assume that these slots are rectangular in shape. The distance between the slots is T, and the width of slots itself equals τ_0 in time domain. The ratio $K = \tau_0 / T$ represents the length of disturbance as a part of impulse repetition. The set of slots during cutting may be considered as random impulse process with mean height value *a* and height variance σ^2 for it. The spectrum of this process $G_{zz}(f)$

$$G_{zz}(f) = [4\tau_0^2/T]^*\{\sin^2(\pi f\tau_0) / (\pi f\tau_0)^2\}^*$$

$$\{\sigma^2 + (2\pi/T)\, a^2 \sum_{i=0}^{\infty} \delta[2\pi(f - i/T)] \qquad (18)$$

The first term in equation (18) represent the continuos part of the spectrum, and the second - the discrete one. If the height of slots is constant, then the first term in equation (18) equals 0. The disturbance spectrum in this case will be presented by set of spectral harmonics lines (at frequencies i/T, where i - harmonic #) , as it is shown in Figure 4. The amplitude of this lines is modulated by the curve $[\sin^2(\pi f\tau_0) / (\pi f\tau_0)^2]$. The interval between them Δf equals 1/ T.

Let's assume that FRF has to be evaluated in the frequency region with maximum frequency f_m. As it follows from Figure 4, the disturbance spectrum equals 0 at frequency $f_{\tau 0} = 1/\tau_0$. The disturbance will have enough energy if $f_m \leq f_{\tau 0}/2$. The number of spectrum lines N_l in frequency interval $(0, f_m)$ equals $f_m/\Delta f$

$$N_l = f_m / \Delta f = T/(2\tau_0) = 0.5/K \qquad (19)$$

Presented above relations are useful for test cutting parameters calculation.

Example

The dynamic stiffness of the lathe with a ~~turret was~~ evaluated. The goal of the test (17) was to estimate the frequency response function (FRF) of the system in the range 60-500 Hz.

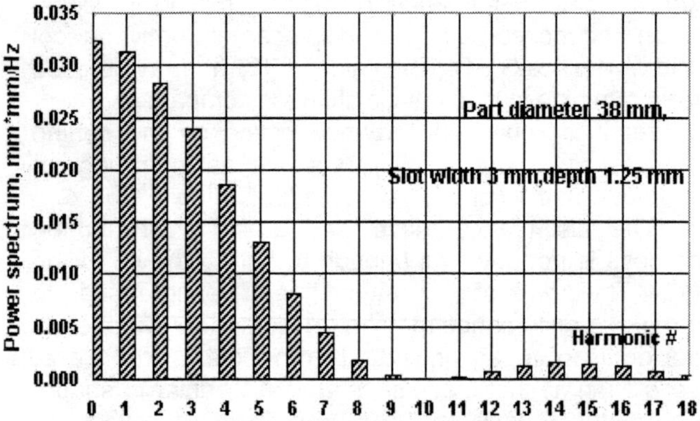

Fig.4. Disturbance spectrum of a workpiece, used in lathe dynamic compliance eveluation during cutting.

The number of spectrum lines has to be used in analysis $N_l = 5$.

The ratio $K = \tau_0 / T = 0.5/N_l = 0.1$

From the other side $f_{\tau 0} = 2\ f_m = 2*500 = 1000$ Hz, and $\tau_0 = 1/1000 = 0.001$ sec.

The distance between the slots in time domain $T = \tau_0 /0.1 = 0.001/0.1 = 0.01$ sec. The defined time domain parameters T and τ_0 allows us to calculate cutting parameters.

The diameter of the part in the test equals 38 mm. The number of slots, that it is reasonable to cut at the part, equals 4. Then the width of the slots w has to be equal

$$w = (3.14*38)*0.1/4 = 3 \text{ mm}$$

Different RPM were used to cover required frequency range for FRF evaluation.

The required height *a* of the slots has to be chosen to minimize the bias error of FRF estimation based on the level of machine tool idle vibration. It may be obtained from expression

$$a \geq (1.5 \div 3) \sqrt{f^* G_{yy} \text{ idle}(f)} \qquad (20)$$

Where G_{yy} idle(f) is a power density of machine idle vibration at the frequency f, at which the $|W_{es}(f)|$ has to be estimated in the test. As it is was found during idle the machine at different RPM the level of vibration did not exceed 0.1 μm and spectrum density of idle vibration do not exceed 0.65e-9 mm^2/Hz. As it follows from equation (20), the height of the slots *a* has to be larger than 2 μm. From a practical point of view we choose height of the slots to equal 1.25 mm. As it follows from Figure 4 the density G_{zz} of excitement at harmonic number 5 will be equal 0.013 mm^2/Hz. The test was performed at different RPM with depth of cut 2.5 mm.

During the test, the spectrum analyzer measured the amplitude spectrum of topplate and spindle headstock vibration. The amplitude of vibration at all exciting harmonics, presented in the spectrum, was recorded. The density spectrum was calculated based at these experimental data for every harmonic.

The dynamic compliance data, obtained at every frequency, at different RPM, was averaged. The obtained FRF is presented in Figure 5.

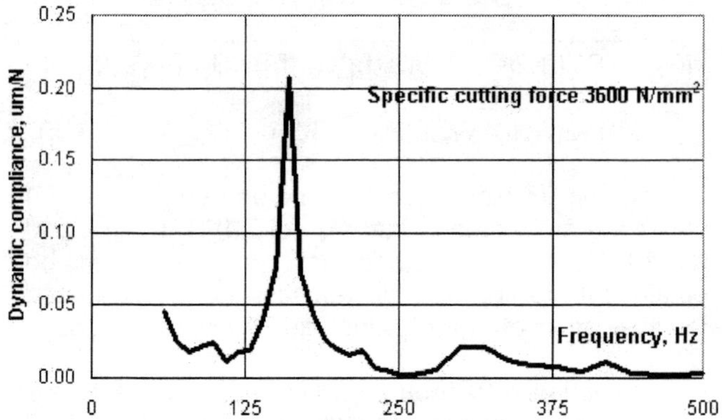

Fig.5. FRF of lathe , obtained during cutting.

The bias error estimation was based on obtained experimental data. The ratio G_{yy} idle (f)/G_{yy}(f) did not exceed 1e-8, and $|W_c(f)^* W_{es}(f)|$ did not exceed 4e-3.

The hammer test was performed to obtained data for comparison. The topplate was loaded by impact in the X direction and vibration of the topplate were measured in the same point that during cutting. The

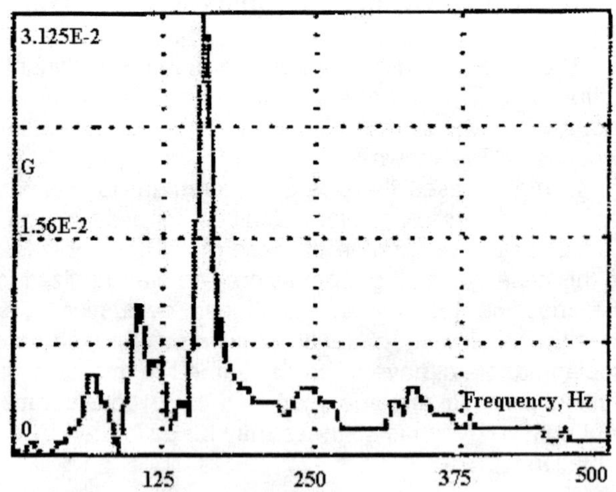

Fig.6. Spectrum of machine structure vibration under impact load.

obtained vibration spectrum of topplate acceleration under impact load is presented at Figure 6.

The level of vibration of headstock in the same frequency range under similar load was more than 10 times less.

The curve at Figure 5 explicitly shows that turret-crosslide-carriage assembly has a natural mode of vibration at 160 Hz. Others natural modes excited during cutting are in the range of 100, 320, and 440 Hz.

It is obvious from comparison of curves at Figure 5 and Figure 6 that 160 Hz is a natural frequency of the tested mechanical system. The same conclusion can be made about the region near 100 Hz. Both natural modes were excited during cutting and during impact test. The modes at 320 and 440 Hz were not excited during impact test. From the other side, the modes at 62 and 240 Hz were not excited during cutting.

These differences in vibration spectrum represent the basic difference in the data obtained under an artificial load and in real cutting condition. This difference has to be taken in the account during machine tool dynamic testing set ups and interpretation of obtained data.

CONCLUSION.

1. The theoretical analysis based at the theory of stationary random processes, shows that dynamic compliance of a machine tool structure may be evaluated during cutting.
2. Defined cutting test conditions, which may provide evaluation of machine tool frequency response function in a wide frequency range with a reasonable accuracy.
3. A simplified approach is proposed, in which only relative vibration between tool and workpiece are measured. This approach may be used for experimental dynamic stiffness evaluation during cutting and machine quality comparison.
4. Cutting tests results are presented, which validate practical application of proposed methodology

REFERENCES

Kaminskaya V.V, Kushnir E.F. , 1979, " Dynamic features of Metal-Cutting process with free chip formation", Machine & Tooling, Volume 50, No 5, pp. 34-38

Moriwaki, T.,Ivata, K. 1976, "Analysis of the dynamic characteristics of machine tool structures in cutting", Transactions ASME, vol. 55, No. 4., pp. 256-262.??

Bendat J.S., Piersol A.G., 1971. Random data. Analysis and measurement procedures. Wiley-Interscience. ISBN 0-471-06470-X, New York, , 407 pp.

J. Tlusty, T. Moriwaki, 1976 "Experimental and Computational Identification of Dynamic Structural Models"' CIRP Annals, Vol. 25/2.

Proceedings of IMECE04
2004 ASME International Mechanical Engineering Congress and Exposition
November 13-20, 2004, Anaheim, California USA

IMECE2004-59766

"FREE ASYMMETRIC VIBRATIONS OF COMPOSITE FULL CIRCULAR CYLINDRICAL SHELLS STIFFENED BY A BONDED CENTRAL SHELL SEGMENT"

by

U. Yuceoglu#
Department of Aerospace Engineering
Middle East Technical University
Ankara, 06531, Turkey .

Phone: (++90) - (312)-210 24 71, 210 42 84
FAX: (++90) - (312) - 210 12 72
e-mail: yuceoglu@ae.metu.edu.tr

V. Özerciyes*
Aero-Structural Design Department
TUSAŞ Aerospace Industries, Inc. (TAI)
Ankara, 06531, Turkey .

Phone: (++90) - (312)-210 1389
FAX: (++90) - (312) - 210 12 72
e-mail: vozerciyes@tai.com.tr

ABSTRACT

This study is concerned with the "Free Asymmetric Vibrations of Composite Full Circular Cylindrical Shells Stiffened by a Bonded Central Shell Segment". The base shell is made of an orthotropic "full" circular cylindrical shell reinforced and/or stiffened by an adhesively bonded dissimilar, orthotropic "full" circular cylindrical shell segment. The stiffening shell segment is located at the mid-center of the composite system. The theoretical analysis is based on the "Timoshenko-Mindlin-(and Reissner) Shell Theory" which is a "First Order Shear Deformation Shell Theory (FSDST)". Thus, in both "base (or lower) shell" and in the "upper shell" segment, the transverse shear deformations and the extensional, translational and the rotary moments of inertia are taken into account in the formulation. In the very thin and linearly elastic adhesive layer, the transverse normal and shear stresses are accounted for. The sets of the dynamic equations, stress-resultant-displacement equations for both shells and the in-between adhesive layer are combined and manipulated and are finally reduced into a "Governing System of the First Order Ordinary Differential Equations" in the "state-vector" form. This system is integrated by the "Modified Transfer Matrix Method (with Chebyshev Polynomials)". Some asymmetric mode shapes and the corresponding natural frequencies showing the effect of the "hard" and the "soft" adhesive cases are presented. Also, the parametric study of the "overlap length" (or the bonded joint length) on the natural frequencies in several modes is considered and plotted.

Professor, Member-ASME, * Ph. D.

INTRODUCTION

In flight vehicle structures such as aircraft fuselages, rocket motor cases, booster vehicle structures, etc. the "Advanced Metal Alloy Stiffened Cylindrical Shells" and, currently and more increasingly, the "Advanced Composite Stiffened Cylindrical Shells" are being used [1,2,3]. The "Advanced Composites" are also used in strengthening and/or repairing metal alloy or composite cylindrical shell and other aero-structural systems by means of the adhesively bonded shell strips or patches, or "full" cylindrical shell segments [2,3]. In this way, the dynamic response of the aero-structures or systems is reduced to an acceptable and safe level. Thus, the dynamic response is modified and the integrity of the structure is maintained. Also, if there are any crack (or cracks) in the structure, an adhesively bonded, composite "shallow" shell strip or a "full" cylindrical shell segment can eliminate further development and propagation of existing crack(s) without significantly increasing the dead-weight of the aero-structure [4].

The "Advanced Metal Alloy Stiffened Cylindrical Shells" may be categorized in terms of the stiffener (transverse) types:

(1) Circular Cylindrical Shells with a Single Central or Non-Central Ring Stiffener
(2) Circular Cylindrical Shells with Equidistant (or Periodic) Ring Stiffeners
(3) Circular Cylindrical Shells with Arbitrarily Spaced (Non-Equidistant) Ring Stiffeners
(4) Circular Cylindrical Shells with Grid-type (or Orthogonal) Stiffeners

The literature on the "Advanced Metal Alloy Stiffened Cylindrical Shells" are exhaustive. In this connection, one may mention a recent review study (up to 1992) of stiffened shells with above mentioned stiffener types by Mukhopadhyay and Sinha [5].

In the "Advanced Composite Stiffened Cylindrical Shells", a similar categorization may be assumed in terms of the stiffener (transverse) types. In this study, however, the transverse stiffener(s) is (are) assumed as in the form of "full" shell segment(s) adhesively bonded to the base (or main) cylindrical shell.

Some recent investigations on the vibrations of composite stiffened shallow cylindrical shells are presented in Yuceoglu and Özerciyes [6-8]. In the present study, the composite stiffened "full" circular cylindrical shells of type (1) are to be analyzed. Any studies on the free vibrations of type (1) composite "full" cylindrical shell system, to the best knowledge of the present authors, are not available (or very few, if any) in the open literature.

Therefore, the main purpose of this study is to formulate and to investigate the "Free Asymmetric Vibrations of Composite Full Circular Cylindrical Shells Stiffened by a Bonded Central Shell Segment".

In the analytical formulation, the "Timoshenko-Mindlin-(and Reissner) Shell Theory" [9,10] which is a "First Order Shear Deformation Shell Theory (FSDST)" is used [9-13]. Hence, in the analyses of the shell layers, their masses and their extensional, translational and rotary moments of inertia as well as the transverse shear deformations of shells will be included. Since this is a preliminary study of the problem under consideration, only a "(FSDST)" has been employed instead of one of the several more complicated "Higher Order Shear Deformation Shell Theories (HSDST)" available in the literature such as Reddy and Liu [14].

THEORETICAL ANALYSIS AND REDUCTION OF SYSTEM OF EQUATIONS

The general configuration and the characteristic geometry of the "Composite Full Cylindrical Shells Stiffened by a Bonded Central Shell Segment" are presented in Figure 1.a and the longitudinal cross-section of the entire assembly is shown in Figure 1.b. In the analytical formulation, the base (or lower) "full" circular cylindrical shell and the upper stiffening "full" shell segment are considered as dissimilar orthotropic shells connected by an adhesive layer. . In the relatively very thin and linearly elastic adhesive layer, due to its very small thickness, the mass and the moment of inertia are neglected. However, adhesive layer transverse normal and shear stresses on the shell-adhesive layer interfaces are included in the analysis.

The coordinate systems are attached to the median surfaces of shells and the orthotropic material directions are indicated in Figures 1.a and 1.b. In both the base (or lower) shell and the upper stiffening shell segment, the transverse shear deformations, the masses and the translational, extensional and the rotary moments of inertia are to be included in the formulation. Thus, the present analysis is based on the well-known "Timoshenko-Mindlin (and Reissner)" type shell theory which is a "First Order Shear Deformation Shell Theory (FSDST)" [9-13].

The sign convention for the stress resultants is based on the classical elasticity theory of sign convention for stresses. The more detailed theoretical developments can be found in a recent work by the present authors Yuceoglu and Özerciyes [15-17]. Therefore, in this study, these details will be omitted.

In order to ease the present theoretical formulation and the method of solution, the entire composite, stiffened shell system is divided into three parts or regions in the longitudinal direction, as Part I, Part II and Part III. (see also Figure 1.b). As a beginning step, the dynamic equations of the two shells and the adhesive layer, the stress resultant-displacement relations and the constitutive (Hooke's Law) equations of orthotropic shells given in Appendix A are combined in a "special form". Thus, for a single layer "full" circular cylindrical shell, this "special form" is as follows:

$$\left|\begin{aligned}
&\frac{\partial u}{\partial x}=\frac{1}{hB_{11}}N_{xx}-\frac{B_{12}}{B_{11}}\left[\frac{1}{R}\frac{\partial v}{\partial \varphi}+\frac{w}{R}\right]\\
&\frac{\partial v}{\partial x}=\frac{1}{hB_{66}}N_{x\psi}-\frac{1}{R}\frac{\partial u}{\partial \varphi}\\
&\frac{\partial w}{\partial x}=\frac{1}{\kappa_x^2 hB_{55}}Q_{xz}-\beta_x\\
&\frac{\partial \beta_x}{\partial x}=\frac{12}{h^3B_{11}}M_{xx}-\frac{B_{12}}{B_{11}}\frac{1}{R}\frac{\partial \beta_\varphi}{\partial \varphi}\\
&\frac{\partial \beta_\varphi}{\partial x}=\frac{12}{h^3B_{66}}M_{x\varphi}-\frac{1}{R}\frac{\partial \beta_x}{\partial \varphi}
\end{aligned}\right. \tag{1.a}$$

$$\left|\begin{aligned}
&\frac{\partial N_{xx}}{\partial x}=-\frac{1}{R}\frac{\partial N_{x\varphi}}{\partial \varphi}-p_x+\rho h\ddot{u}\\
&\frac{\partial N_{x\varphi}}{\partial x}=-\frac{B_{12}}{B_{11}}\frac{1}{R}\frac{\partial N_{xx}}{\partial \varphi}+\frac{h}{R^2}\frac{B_{12}^2}{B_{11}}\left[\frac{\partial^2 v}{\partial \varphi^2}+\frac{\partial w}{\partial \varphi}\right]\\
&\qquad-\frac{hB_{22}}{R^2}\left[\frac{\partial^2 v}{\partial \varphi^2}+\frac{\partial w}{\partial \varphi}\right]\\
&\qquad-\frac{\kappa_\varphi^2 hB_{44}}{R^2}\left[\beta_\varphi-\frac{v}{R}+\frac{1}{R}\frac{\partial w}{\partial \varphi}\right]-p_\varphi+\rho h\ddot{v}
\end{aligned}\right.$$

$$\frac{\partial M_{xx}}{\partial x} = -\frac{1}{R}\frac{\partial M_{x\varphi}}{\partial \varphi} + Q_{xz} - m_{\varphi} + \rho\frac{h^3}{12}\ddot{\beta}_x \quad (1.b)$$

$$\frac{\partial M_{x\varphi}}{\partial x} = -\frac{B_{12}}{B_{11}}\frac{1}{R}\frac{\partial M_{xx}}{\partial \varphi} + \frac{h^3}{12}\frac{B_{12}^2}{R^2 B_{11}}\frac{\partial^2 \beta_{\varphi}}{\partial \varphi^2} - \frac{h^3}{12}\frac{B_{22}}{R^2}\frac{\partial^2 \beta_{\varphi}}{\partial \varphi^2} + \kappa_{\varphi}^2 h B_{44}\left[\beta_{\varphi} - \frac{v}{R} + \frac{1}{R}\frac{\partial w}{\partial \varphi}\right] - m_x + \rho\frac{h^3}{12}\ddot{\beta}_{\varphi}$$

$$\frac{\partial Q_{xz}}{\partial x} = -\kappa_{\varphi}^2 h B_{44}\left[\frac{1}{R}\frac{\partial \beta_{\varphi}}{\partial \varphi} - \frac{1}{R^2}\frac{\partial v}{\partial \varphi} + \frac{1}{R^2}\frac{\partial^2 w}{\partial \varphi^2}\right] + \frac{B_{12}}{B_{11}}\frac{N_{xx}}{R} - \frac{h}{R^2}\frac{B_{12}^2}{B_{11}}\left[\frac{\partial v}{\partial \varphi} + w\right] + \frac{hB_{22}}{R^2}\left[\frac{\partial v}{\partial \varphi} + w\right] - p_z + \rho h\ddot{w}$$

In the above equations "**p**" and "**m**" indicate the surface loads and moments on the reference surface due to the adhesive stresses acting on the upper/lower surfaces of shells.

The above set of equations can be rewritten in a compact "matrix form" for Part I, Part II and Part III, respectively:

In Part I region ("Overlap Region" with two-layer shell and in-between adhesive layer):

$$\frac{\partial}{\partial x_I}\begin{Bmatrix} \mathcal{Z}^{(1)} \\ \mathcal{Z}^{(2)} \end{Bmatrix} = [\mathcal{E}]\begin{Bmatrix} \mathcal{Z}^{(1)} \\ \mathcal{Z}^{(2)} \end{Bmatrix} \qquad (0<x_I<\ell_I) \qquad (2)$$

In Part II region (single layer base shell):

$$\frac{\partial}{\partial x_{II}}\{\mathcal{Z}^{(2)}\} = [\mathcal{F}]\{\mathcal{Z}^{(2)}\} \qquad (0<x_{II}\leq \ell_{II}) \qquad (3)$$

In Part III region (single layer base shell),

$$\frac{\partial}{\partial x_{III}}\{\mathcal{Z}^{(2)}\} = [\mathcal{G}]\{\mathcal{Z}^{(2)}\} \qquad (0\leq x_{III}<\ell_{III}) \qquad (4)$$

where the "fundamental dependent variables" of the problem can be written in a column matrix form or in the "state vector" form as $\{\mathcal{Z}^{(i)}\}$ for the upper (i=1) and lower (i=2) "full" circular cylindrical shell elements,

$$\{\mathcal{Z}^{(i)}\} = \begin{Bmatrix} u^{(i)}, v^{(i)}, w^{(i)}, \beta_x^{(i)}, \beta_{\varphi}^{(i)}; \\ N_{xx}^{(i)}, N_{x\varphi}^{(i)}, M_{xx}^{(i)}, M_{x\varphi}^{(i)}, Q_{xz}^{(i)} \end{Bmatrix}^T \qquad (i=1,2) \qquad (5)$$

which includes the displacements, angles of rotation and the stress resultants appearing at any x=const. of the composite stiffened shell system.

At this point, the non-dimensionalizations of the equations in (2), (3), (4) are to be considered. Thus, ρ_2, h_2, R_2, $B_{11}^{(2)}$ are chosen as the reference quantities or parameters. Then, the dimensionless quantities are:

The dimensionless coordinates in the x-direction,

$$\xi_I = x_I/\ell_I \quad \text{(in Part I. region)}$$
$$\xi_{II} = x_{II}/\ell_{II} \quad \text{(in Part II. region)} \qquad (6)$$
$$\xi_{III} = x_{III}/\ell_{III} \quad \text{(in Part III. region)}$$

The dimensionless parameters related to the densities and geometry of the shells and the adhesive layer,

$$\bar{\rho} = \rho_1/\rho_2, \quad \bar{h}_a = h_a/h_2, \quad \bar{R} = R_1/R_2, \quad \bar{h} = h_1/h_2$$

$$\bar{\ell}_k = \frac{\ell_k}{R_2} \quad \textbf{(k=I, II and III for Parts or Regions)} \qquad (7)$$

The dimensionless parameters involving the "full" shell elements and the adhesive layer elastic constants,

$$\bar{B}_{jk}^{(i)} = B_{jk}^{(i)}/B_{11}^{(2)} \quad (i=1,2 \text{ and } j,k=1,2),$$
$$\bar{B}_{\ell\ell}^{(i)} = B_{\ell\ell}^{(i)}/B_{11}^{(2)} \quad (i=1,2 \text{ and } \ell=4,5,6) \qquad (8)$$
$$\bar{G}_a = G_a/B_{11}^{(2)}, \quad \bar{E}_a = E_a/B_{11}^{(2)}$$

The dimensionless natural frequency (or the circular frequency) of the entire composite shell system,

$$\bar{\omega}_{mn} = \frac{\rho_2 R^{7/2}\omega_{mn}^2}{h_2^{3/2} B_{11}^{(2)}} \qquad (m,n=1,2,3...) \qquad (9)$$
$$\bar{\Omega} = \bar{\omega}_{mn}$$

The independent space variable "φ" and the time "t" can be eliminated from the equations. Then, by making use of the "periodicity" and the "continuity" of the "dependent fundamental variables", the Trigonometric Fourier Expansions can be obtained in the φ-direction. Then, the generalized displacements can be expressed as (B.1) in Appendix B. Similarly, b

The simple insertion of (B.1) into the differential equations of (2), (3), (4) respectively yields a reduced system of equations. The dimensionless set of the "Governing System of the First Order Ordinary Differential Equations" in the "state vector" form are:

In Part. I region (two-layer shell and in-between adhesive layer):

$$\frac{d}{d\xi_I}\begin{Bmatrix} \overline{\mathcal{Z}}_{mn}^{(1)} \\ \overline{\mathcal{Z}}_{mn}^{(2)} \end{Bmatrix} = \begin{bmatrix} \overline{\mathcal{E}}_{11} & \overline{\mathcal{E}}_{12} \\ \overline{\mathcal{E}}_{21} & \overline{\mathcal{E}}_{22} \end{bmatrix} \begin{Bmatrix} \overline{\mathcal{Z}}_{mn}^{(1)} \\ \overline{\mathcal{Z}}_{mn}^{(2)} \end{Bmatrix},$$

$(0 < \xi_I < 1)$ (upper stiffening shell segment)
$(0 \le \xi_I \le 1)$ (base or lower shell) (10)

with the appropriate "continuity conditions" and the "boundary conditions".

In Part. II region (single layer shell):

$$\frac{d}{d\xi_{II}}\left\{\overline{\mathcal{Z}}_{mn}^{(2)}\right\} = \left[\mathcal{F}\right]\left\{\overline{\mathcal{Z}}_{mn}^{(2)}\right\},$$

$(0 < \xi_{II} \le 1)$ (base or lower shell) (11)

with the appropriate "continuity conditions" and the "boundary conditions".

In Part. III region (single layer shell):

$$\frac{d}{d\xi_{III}}\left\{\overline{\mathcal{Z}}_{mn}^{(2)}\right\} = \left[\mathcal{G}\right]\left\{\overline{\mathcal{Z}}_{mn}^{(2)}\right\},$$

$(0 \le \xi_{III} < 1)$ (base or lower shell) (12)

with the appropriate "continuity conditions" and the "boundary conditions".

where the dimensionless column matrix or the "state vector" $\left\{\overline{\mathcal{Z}}_{mn}^{(i)}\right\}$ is now,

$$\left\{\overline{\mathcal{Z}}_{mn}^{(i)}\right\} = \begin{Bmatrix} \overline{U}_{mn}^{(i)}, \overline{V}_{mn}^{(i)}, \overline{W}_{mn}^{(i)}, \overline{\beta}_{mnx}^{(i)}, \overline{\beta}_{mn\varphi}^{(i)}; \\ \overline{N}_{mnxx}^{(i)}, \overline{N}_{mnx\varphi}^{(i)}, \overline{M}_{mnxx}^{(i)}, \overline{M}_{mnx\varphi}^{(i)}, \overline{Q}_{mnxz}^{(i)} \end{Bmatrix}^T$$

(i=1,2) (13)

which includes all the "dimensionless fundamental dependent variables" of the problem.

The first order system in (10), (11), (12) constitutes a "Two-Point Boundary Value Problem". Since the coefficient matrices in the above equations include $\overline{\omega}_{mn}$, these equations can be integrated numerically once the dimensionless natural frequency $\overline{\omega}_{mn}$ of the entire composite shell system is obtained or known for a given case.

SOLUTION METHOD AND PROCEDURE

The governing system of equations in (10), (11) and (12) can be integrated by means of the "Modified Transfer Matrix (with Chebyshev Polynomials)" [15-18] developed by the present authors. This semi-analytical and numerical procedure is a combination of the "Classical Lévy's Method", the "Transfer Matrix Method", and the "Integrating Matrix Method (with Chebyshev Polynomials)". The present solution technique is a further extension of the method of solution employed in a certain class of shallow shell free vibration problems by Yuceoglu and Özerciyes [6-8]. A similar numerical solution technique in some stiffened plate vibrations problems is used in Yuceoglu and Özerciyes [19]. An earlier version for some plate lap joint and other plate vibration problems is used in Yuceoglu et al. [20]. It would be of interest to point out here that the "Modified Transfer Matrix Method (with Interpolation Polynomials or with Chebyshev Polynomials)" is a very efficient and accurate technique based on the experience of the present authors in solving a certain class of plate and shell vibration problems.

Here, the main steps of the present solution procedure will be briefly explained. As a first step in the method, the governing system of differential equations given in (10), (11), (12) are discretized (see Eq. (B.3) in Appendix B). This is done by discretizing the "dimensionless fundamental dependent variables" and the coefficient matrices in those equations with respect to ξ_I, ξ_{II}, ξ_{III}, respectively. Hence, Part I, Part II and Part III regions are divided into sufficient number of segments or stations along the ξ_I, ξ_{II}, ξ_{III} directions.

After this step, the governing differential equations are premultiplied by the appropriate "Global Integrating Matrices" $[\mathcal{L}]$ which includes 20 square "Integrating Sub-Matrices $[\ell]$" of dimensions ($20k_1 x 20k_1$). Here, "k_1" is the number of discretizing points along the ξ_I-direction in Part I region. Thus, these operations result in the following system:

In Part I region (two-layer full shell and adhesive layer),

$$\begin{Bmatrix} \dot{\overline{\mathcal{Z}}}_{mn}^{(1)} \\ \dot{\overline{\mathcal{Z}}}_{mn}^{(2)} \end{Bmatrix} - \begin{Bmatrix} \dot{\overline{\mathcal{Z}}}_{\xi_I=0}^{(1)} \\ \dot{\overline{\mathcal{Z}}}_{\xi_I=0}^{(2)} \end{Bmatrix} = \left[\mathcal{L}\right]\left[\dot{\overline{\mathcal{E}}}\right] \begin{Bmatrix} \dot{\overline{\mathcal{Z}}}_{mn}^{(1)} \\ \dot{\overline{\mathcal{Z}}}_{mn}^{(2)} \end{Bmatrix}, (0 \le \xi_I \le 1)$$

(14)

where "•" indicates the discretization of a particular matrix along the ξ_I-direction and the "barred" or "--" quantities show the dimensionless terms. The subscript "(ξ_I=0)" in the state vectors in (14) means that the "state vectors" are evaluated at the "initial end point ξ_I =0" in Part I region.

The above equation of (14) can be rearranged in such a way that a relation between the "state vector" at a "general station ξ_I" and a "state vector" at the "initial end point ξ_I =0" is obtained. Then,

$$\begin{Bmatrix} \dot{\overline{\mathcal{Z}}}_{mn}^{(1)} \\ \dot{\overline{\mathcal{Z}}}_{mn}^{(2)} \end{Bmatrix} = \left[\dot{\mathcal{T}}\right] \begin{Bmatrix} \dot{\overline{\mathcal{Z}}}_{\xi_I=0}^{(1)} \\ \dot{\overline{\mathcal{Z}}}_{\xi_I=0}^{(2)} \end{Bmatrix}, \quad \left[\dot{\mathcal{T}}\right] = \left([I] - \left[\mathcal{L}\right]\left[\dot{\overline{\mathcal{E}}}\right]\right)^{-1}$$

(15)

where **[I]** is the unit matrix and [$\dot{\mathcal{T}}$] is the discretized version of the so-called "Modified Transfer Matrix" along the ξ_I-direction between a "general or arbitrary station" and the "initial end point $\xi_I=0$". The discretized "Modified Transfer Matrix" in the above includes the "Integrating Matrices" (see also Appendix B). Thus, in Part. I region, going from station to station along ξ_I, one can obtain a final relation between the "state vectors" at the "initial end point $\xi_I=0$" and the "final end point $\xi_I=1$". In the "Integrating Matrices", the "Chebyshev Polynomials" are employed. For simplification and convenience, the subscript "mn" in (15) are dropped such that,

<u>In Part I region</u> (Two-layer "Full" Shell and Adhesive Layer):

$$\begin{Bmatrix} \dot{\mathcal{Z}}^{(1)}_{\xi_1=1} \\ \dot{\mathcal{Z}}^{(2)}_{\xi_1=1} \end{Bmatrix} = \left[\tilde{\mathcal{T}}\right]_{01} \begin{Bmatrix} \dot{\mathcal{Z}}^{(1)}_{\xi_1=0} \\ \dot{\mathcal{Z}}^{(2)}_{\xi_1=0} \end{Bmatrix}, \quad \left[\tilde{\mathcal{T}}\right]_{01} = \begin{bmatrix} \tilde{\mathcal{U}}_{1,1} & \tilde{\mathcal{U}}_{1,2} \\ \tilde{\mathcal{U}}_{2,1} & \tilde{\mathcal{U}}_{2,2} \end{bmatrix}_{01}$$

$$(0 \le \xi_I \le 1) \qquad (16)$$

where $\left[\tilde{\mathcal{T}}_{i,j}\right]_{01}$ (i,j=1,2) are partioned square matrices of (10x10) which explicitly include the dimensionless natural frequency parameter $\overline{\omega}_{mn}$. In (16), the subscript "01" means that the final form of the "Modified Transfer Matrix" $\left[\tilde{\mathcal{T}}\right]_{01}$ transfers the "state vector" from the "initial end point $\xi_I=0$" (or the initial station) to the "final end point $\xi_I=1$" (or the final station). The end result of the discretization procedure of the "state vectors", the "Coefficient Matrices" and the final form of the "Modified Transfer Matrix", as shown in (16), are given in detail in Yuceoglu and Özerciyes [15-17].

In a similar way, the expressions corresponding to the appropriate final form of the "Modified Transfer Matrices" can be obtained for Part II and Part III regions, in terms of $\left[\tilde{\mathcal{S}}\right]_{01}$ and $\left[\tilde{\mathcal{P}}\right]_{01}$, respectively. After substituting the "continuity conditions" between Part I and Part II, similarly between Part I and Part III regions (see also Figure 1.b) and inserting the "support conditions" in the x-direction of the upperand the lower full shell elements, a matrix equation is obtained:

$$[C_0(\overline{\omega}_{mn})]\{\mathcal{Z}_0\} = \{0\} \qquad (17)$$

where the coefficient matrix **[C$_0$]** includes dimensionless natural frequecies $\overline{\omega}_{mn}$. Therefore, the determinant of the coefficient matrix yields the polynomial which has the natural frequencies as its roots. Thus, the determinant is,

$$|C_0(\overline{\omega}_{mn})| = 0 \qquad (18)$$

The roots or the natural frequencies are,

$$\begin{vmatrix} \overline{\Omega} = \overline{\omega}_{mn} \quad (m,n = 1,2,3...) \\ \overline{\Omega}_1 < \overline{\Omega}_2 < \overline{\Omega}_3 < .... \end{vmatrix} \qquad (19)$$

where the dimensionless natural frequencies $\overline{\Omega}$'s in (19) are sequenced or ordered according to their magnitudes as shown above. Depending on the given "n" and the assigned "m" values and their magnitudes, the sequencing of the $\overline{\Omega}_1, \overline{\Omega}_2, \overline{\Omega}_3, ...$ indicate first, second, third,..., natural frequencies, respectively. Here, "m " is a dummy subscript, while "n" is coming from the trigonometric series or rather from the "Classical Levy's Solution". (see also Appendix B).

SOME NUMERICAL APPLICATIONS and BRIEF CONCLUSIONS

The present theoretical analysis and formulation and the solution procedure are applied to the problem under consideration here. Hence, some mode shapes with their corresponding natural frequencies, and the numerical results of a few typical parametric study cases will be presented.

The geometric and material characteristics of the composite stiffened "full" cylindrical shell system are given in Table. 1. It should be noted that, in Table. 1, the "hard" adhesive is from commercially available adhesives. However, the "soft" adhesive is somewhat fictious and is used only in computations in order to simulate and to compare the numerical results with those of the realistic "hard" adhesive case.

The boundary conditions on the Figures are the support conditions in the x-direction only. They read from left to right. Hence, the first two letters indicate the boundary conditions for the upper "full" shell segment and the subsequent two letters show the support conditions for the lower (or base) "full" circular cylindrical shell all in the x-direction. (See also Appendix A.)

In what follows, some typical mode shapes and a few parametric studies on natural frequencies will be presented. In Figure 2, in the hard adhesive case, the first three consequtive mode shapes and the corresponding dimensionless natural frequencies are shown for the (FFCC) boundary conditions in the x-direction. Similarly, in Figure 3, in the "soft" adhesive case, again the first three mode shapes with the corresponding natural frequencies are shown for the same support conditions. One can easily observe that, by a simple comparison, the mode shapes (as well as the frequencies) associated with each case are quite different from each other. Also, in the "hard" adhesive case, the bonded central stiffening shell segment moves with the base (or lower) "full" cylindrical shell. This is, of course, due to the relatively high stiffness of the "hard" adhesive. In the "soft" adhesive case, in Figure 3, it can be clearly seen from the mode shapes that there are significant seperation or delamination areas (i.e. lightly shaded areas in

Figures) in the bonded region in-between the base shell and the stiffening shell segment. This is the consequence of the relatively less stiff character of the "soft" adhesive layer. Hence, one can conclude that the "hard"ness and the "soft"ness of the adhesive layer have serious effects on the mode shapes, the natural frequencies and on the delamination failure of the bonded or the overlap region. This, of course, may result in a possible failure mode of the entire composite, stiffened "full" cylindrical shell system.

In Figures 4 and 5, the natural frequencies vs. "Joint Length Ratio ℓ_I/L " for the first four modes are plotted for the same (FFCC) support conditions in the "hard" and the "soft" adhesive cases, respectively. In the "soft" adhesive case in Figure 5, it can be seen that the natural frequencies gradually decrease while the "Joint Length Ratio ℓ_I/L " increases, whereas in Figure 4, in the "hard" adhesive case, no such pattern can be observed.

The effect of the "Bending Rigidity Ratio $D_{11}^{(1)}/D_{11}^{(2)}$ " on the first four natural frequencies for the same (FFCC) boundary conditions is also considered. For this purpose, Figures 6 and 7, corresponding to the "hard" and the "soft" adhesive cases, respectively are presented. It is very interesting to observe that, in both Figures, the natural frequencies remain almost constant as the "Bending Stiffness ratio $D_{11}^{(1)}/D_{11}^{(2)}$ " increases. In otherwords, in both "hard" and "soft" adhesive cases, the relative increases in the bending stiffness of the stiffening shell segment have no significant (or very little) effect on the natural frequencies of the entire composite stiffened shell system in all four modes. Although, it is not shown here, similar observations are made for other support conditions. If the support conditions are less constraining such as (FFCF), etc., then only the fourth natural frequency in the "hard" adhesive case show some gradual increases. Otherwise, for both "hard" and "soft" adhesive cases, the natural freqıencies have very slight or no increases while the "Bending Stiffness ratio $D_{11}^{(1)}/D_{11}^{(2)}$ " increases.

REFERENCES

[1] Hoskins, B.C. and Baker, A.A., 1986, "Composite Materials for Aircraft Structures", AIAA Educational Series, New York.

[2] Marshall, I.H., and Demuts, E. (Editors), 1988, "Supportability of Composite Airframes and Structures", Elsevier Applied Science Publishers, New York.

[3] Smith, P.J., Ilcewicz, L.B., and Olson, Y. T., 1995, "Advanced Technology Composite Fuselage", Proceedings of the 5th NASA/DOD/ARPA Advanced Composites Technology Conference, NASA-Langley Research Center, (also NASA CP-3294), pp.1-31.

[4] Baker, A. A. and Jones, R., 1988, "Bonded Repair of Aircraft Structures", Martinus Nijhoff Publishers, Dordrecht, the Netherlands.

[5] Mukhopadyay, M. and Sinha, G., "A Review of Dynamic Behavior of Stiffened Shells" The Shock and Vibration Digest, Vol. 24 , No.8, pp. 3-13.

[6] Yuceoglu, U. and Özerciyes, V. 2001, "Free Vibrations of Composite, Shallow Circular Cylindrical Shell Panels with a Bonded Central Stiffening Shell Strip", in the ASME Aerospace Division "Symposium in Honor of Prof. G. J. Smitses" (2001 ASME Inter. Mech. Engineering Congress & Exposition, Nov. 11-16, 2001, N. Y. City.).

[7] Özerciyes, V. and Yuceoglu, U., 2002, "Free Vibrations Of Composite Shallow Circular Cylindrical Shell Panels Stiffened By A Bonded Non-Central Shell Strip", 43rd AIAA/ASME /ASCE/AHS/ASC/ - "SDM" Conference and Exhibit, Denver, CO., Paper No. AIAA-2002-1505.

[8] Özerciyes, V. and Yuceoglu, U., 2002, "Comparison of Vibration Characteristics of Stiffened Composite Shallow Circular Cylindrical Shells or Panels", in the "Symposium of Recent Advances in Solids and Structures" (2002 ASME Inter. Mech. Engineering Congress & Exposition, Nov. 17-22, 2002, New Orleans, Louisiana).

[9] Soedel, W., 1982, "On the Vibrations of Shells with Timoshenko-Mindlin Type Shear Deflections and Rotatory Inertia", Journal of Sound and Vibration, Vol. 83, pp. 67-79.

[10] Soedel, W. 1993, "Vibrations of Shells and Plates", 2nd Edition, Marcel Dekker, Inc., N.Y.

[11] Mirsky, I., 1964, "Vibrations of Orthotropic, Thick, Cylindrical Shells", Journal of the Acoustical Society of America, Vol. 36, No. 1, pp. 41-51.

[12] Leissa, A. W., 1973, "Vibration of Shells", NASA SP.288.

[13] Markus, S. 1988, "The Mechanics of Vibrations of Cylindrical Shells", Studies in Applied Mechanics, Elsevier Science Publishing Company, N.Y.

[14] Reddy, J. N. and Liu, C. F., 1985, "A Higher Order Shear Deformation Theory of Laminated Elastic Shells", Inter. Jour. Of engineering Science, Vol. 23, No. 3, pp. 319-390.

[15] Yuceoglu, U. and Özerciyes, V., 2004, "Free Non-Axisymmetric Vibrations of Composite Full Circular Shells with a Bonded Non-Central (or Eccentric) Lap Joint", 45th AIAA/ASME/ASCE/AHS/ASC –"SDM" Conference and Exhibit, Palm Springs, California, Paper No. AIAA 2004-1619.

[16] Özerciyes, V. and Yuceoglu, U., 2003, "Free Asymmetric Vibrations of Composite Full Circular Cylindrical Shells with a Bonded Central Lap Joint", in the "Symposium on Recent Advances in Mechanics of Solids and Structures", Paper No. IMECE-2003-43661, 2003 ASME Inter. Mech. Engineering Congress and Exposition, Nov. 15-21, 2003, Washington, D. C.

[17] Yuceoglu, U. and Özerciyes, V., 2003, "Free Vibration Characteristics of Composite Shallow Circular Cylindrical Shells or Panels with a Non-Central Single Lap Joint ", 44rd AIAA/ASME /ASCE/AHS/ASC/ - "SDM" Conference and Exhibit, Norfolk, Virginia, Paper No. AIAA-2003-1775.

[18] Yuceoglu, U. and Özerciyes, V., 2001, "Free Vibrations of Composite Shallow Circular Cylindrical Shell Panels with a Bonded Single Lap Joint", 42nd AIAA/ASME/ASCE/AHS/ ASC/ - "SDM" Conference and Exhibit, Seattle, WA., Paper No. AIAA-2001-1532.

[19] Yuceoglu, U. and Özerciyes, V., 2003, "Free Flexural Vibrations of Orthotropic Composite Base Plates or Panels with a Non-Central Stiffening Plate Strip", ASME Jour. of Vibrations and Acoustics, Vol. 125, pp. 1-16.

[20] Yuceoglu, U., Toghi, F., and Tekinalp, O., 1996, "Free Bending Vibrations of Adhesively Bonded, Orthotropic Plates with a Single Lap Joint", ASME Journal of Vibration and Acoustics, Vol. 118, pp 122-134.

TABLE. 1
(MATERIAL PROPERTIES AND DIMENSIONS)

Graphite - Epoxy (inner or base shell i=2)	*Kevlar - Epoxy (outer or stiffening shell segment i=1)*	*Adhesive Layer (soft)*	*Adhesive Layer (hard)*
E_x =11.71 GPa E_φ =137.8 GPa $G_{x\varphi}$ =5.51 GPa G_{xz} =2.50 GPa $G_{\varphi z}$ =3.00 GPa $\nu_{x\varphi}$ =0.0213 $\nu_{\varphi x}$ = 0.25 ρ_1 = 1.6 gr/cm^3 $h_1=h_2=h$=0.007 m. R = 0.3 m. b_2=L=1.0 m. $\tilde{b}$ = 0.5 m.	E_x = 5.50 Gpa E_φ = 76.0 Gpa $G_{x\varphi}$ =2.10 Gpa G_{xz} =1.50 Gpa $G_{\varphi z}$ =2.00 Gpa $\nu_{x\varphi}$ = 0.024 $\nu_{\varphi x}$ = 0.34 ρ_2 =1.3 gr/cm^3 $h_2=h_1=h$=0.007 m. R = 0.3 m. $b_1=\ell_1$=0.3 m.	$E_a/B_{11}^{(1)} \approx 2.0 \times 10^{-5}$ $G_a/B_{11}^{(1)} \approx 1.0 \times 10^{-5}$ $h_a = 0.15 \times 10^{-3}$ m. ρ_a=neglected - - - - -	$E_a/B_{11}^{(1)} = 3.40 \times 10^{-1}$ $G_a/B_{11}^{(1)} = 1.19 \times 10^{-1}$ $h_a = 0.15 \times 10^{-3}$ m. ρ_a=neglected - - - - - E_a =4.0 GPa G_a=1.4 GPa

APPENDIX A.

DYNAMIC EQUATIONS OF FULL CIRCULAR CYLINDRICAL SHELLS

- **Dynamic Equations ("Timoshenko-Mindlin- (and Reissner) Type Shell Theory (FSDST)"):**

$$\frac{\partial N_{xx}}{\partial x}+\frac{1}{R}\frac{\partial N_{x\varphi}}{\partial \varphi}+p_x=\rho h\ddot{u}$$
$$\frac{\partial N_{x\varphi}}{\partial x}+\frac{1}{R}\frac{\partial N_{\varphi\varphi}}{\partial \varphi}+\frac{Q_{\varphi z}}{R}+p_\varphi=\rho h\ddot{v}$$
$$\frac{\partial Q_{xz}}{\partial x}+\frac{1}{R}\frac{\partial Q_{\varphi z}}{\partial \varphi}-\frac{N_{\varphi\varphi}}{R}+p_z=\rho h\ddot{w} \qquad \text{(A.1)}$$
$$\frac{\partial M_{xx}}{\partial x}+\frac{1}{R}\frac{\partial M_{x\varphi}}{\partial \varphi}-Q_{xz}+m_\varphi=\rho\frac{h^3}{12}\ddot{\beta}_x$$
$$\frac{\partial M_{x\varphi}}{\partial x}+\frac{1}{R}\frac{\partial M_{\varphi\varphi}}{\partial \varphi}-Q_{\varphi z}+m_x=\rho\frac{h^3}{12}\ddot{\beta}_\varphi$$

where **p**'s stand for the external surface loads and ***m***'s are for the moments due to the external surface shear loads, (reduced to the reference surfaces). They correspond to the normal and shear stresses in the adhesive layer (see also (A.9)).

- **Orthotropic Stress-Strain Relations [Hooke's Law]:**

$$[\sigma]=\begin{bmatrix} B_{11} & B_{12} & B_{13} & 0 & 0 & 0\\ B_{12} & B_{22} & B_{23} & 0 & 0 & 0\\ B_{13} & B_{23} & B_{33} & 0 & 0 & 0\\ 0 & 0 & 0 & B_{44} & 0 & 0\\ 0 & 0 & 0 & 0 & B_{55} & 0\\ 0 & 0 & 0 & 0 & 0 & B_{66}\end{bmatrix}[\varepsilon] \qquad \text{(A.2)}$$

where B's are related to material constants in the orthotropic stress-strain relations such that,

$$B_{11}=\frac{E_x}{1-\nu_{xy}\nu_{yx}},\quad B_{44}=G_{yz}$$
$$B_{22}=\frac{E_y}{1-\nu_{yx}\nu_{xy}},\quad B_{55}=G_{xz} \qquad \text{(A.3)}$$
$$B_{12}=B_{21}=\nu_{yx}B_{11}=\nu_{xy}B_{22},B_{66}=G_{xy}$$

- **Stress Resultant - Displacements Relations:**

$$\begin{Bmatrix} N_{xx}\\ N_{\varphi\varphi}\\ N_{x\varphi}\end{Bmatrix}=h\begin{bmatrix} B_{11} & B_{12} & 0\\ B_{12} & B_{22} & 0\\ 0 & 0 & B_{66}\end{bmatrix}\begin{Bmatrix} \frac{\partial u}{\partial x}\\ \frac{1}{R}\frac{\partial v}{\partial \varphi}+\frac{w}{R}\\ \frac{1}{R}\frac{\partial u}{\partial \varphi}+\frac{\partial v}{\partial x}\end{Bmatrix} \qquad \text{(A.4)}$$

$$\begin{Bmatrix} M_{xx}\\ M_{\varphi\varphi}\\ M_{x\varphi}\end{Bmatrix}=\begin{bmatrix} D_{11} & D_{12} & 0\\ D_{12} & D_{22} & 0\\ 0 & 0 & D_{66}\end{bmatrix}\begin{Bmatrix} \frac{\partial \beta_x}{\partial x}\\ \frac{1}{R}\frac{\partial \beta_\varphi}{\partial \varphi}\\ \frac{1}{R}\frac{\partial \beta_x}{\partial \varphi}+\frac{\partial \beta_\varphi}{\partial x}\end{Bmatrix} \qquad \text{(A.5)}$$

$$\begin{Bmatrix} Q_{\varphi z}\\ Q_{xz}\end{Bmatrix}=\begin{bmatrix} A_{44} & 0\\ 0 & A_{55}\end{bmatrix}\begin{Bmatrix} \beta_\varphi-\frac{v}{R}+\frac{1}{R}\frac{\partial w}{\partial \varphi}\\ \beta_x+\frac{\partial w}{\partial x}\end{Bmatrix} \qquad \text{(A.6)}$$

where the shell bending stiffness **D's** and the shell shear stiffness **A's** are given as,

$$D_{ik}=\frac{h^3B_{ik}}{12}\ (i,k=1,2)$$
$$D_{66}=\frac{h^3B_{66}}{12}, \qquad \text{(A.7)}$$
$$A_{44}=\kappa_\varphi^2 hB_{44},\qquad A_{55}=\kappa_x^2 hB_{55}$$

where κ_φ^2 and κ_x^2 are the "Shear Correction Coefficients" in the φ and x directions, respectively.

- **Boundary Conditions**

F (free) : $N_{xx}=N_{x\varphi}=M_{xx}=M_{x\varphi}=Q_{xz}=0$

S (shear diaphragm or simple support without axial constraint) : $v=w=\beta_\varphi=N_{xx}=M_{xx}=0$ (A.8)

C (clamped) : $u=v=w=\beta_x=\beta_\varphi=0$

ADHESIVE LAYER NORMAL AND SHEAR STRESSES (at upper (i=1) and at lower (i=2) shell - adhesive layer interfaces):

$$\tau_{zx}^{(i)}=G_a\gamma_{zx}^{(i)}=\frac{G_a}{h_a}\left[\left(u^{(1)}-\frac{h_1}{2}\beta_x^{(1)}-u^{(2)}-\frac{h_2}{2}\beta_x^{(2)}\right)\right]+\Delta_1$$
$$\tau_{z\varphi}^{(i)}=G_a\gamma_{z\varphi}^{(i)}=\frac{G_a}{h_a}\left[\left(v^{(1)}-\frac{h_1}{2}\beta_\varphi^{(1)}-v^{(2)}-\frac{h_2}{2}\beta_\varphi^{(2)}\right)\right]-\Delta_2$$
$$\sigma_z^{(i)}=E_a\varepsilon_z^{(i)}=\frac{E_a}{h_a}\left(w^{(1)}-w^{(2)}\right)\quad (i=1,2) \qquad \text{(A.9)}$$

where E_a and G_a are the adhesive elastic constants and Δ's are coming from the curvature of shells (the adhesive layer stresses act as surface loads on shells in Part I region or the overlap region in the dynamic equations of A.1).

APPENDIX B.

- **Generalized Displacement and Stress Resultants in "Full" Circular Cylindrical Shells:**

<u>In Part I Region:</u>

$$u^{(i)} = h_2 \sum_{m=1}^{\infty}\sum_{n=1}^{\infty} \overline{U}_{mn}^{(i)}(\xi)\cos(n\varphi)e^{j\omega_{mn}t}$$

$$v^{(i)} = h_2 \sum_{m=1}^{\infty}\sum_{n=1}^{\infty} \overline{V}_{mn}^{(i)}(\xi)\sin(n\varphi)e^{j\omega_{mn}t}$$

$$w^{(i)} = h_2 \sum_{m=1}^{\infty}\sum_{n=1}^{\infty} \overline{W}_{mn}^{(i)}(\xi)\cos(n\varphi)e^{j\omega_{mn}t}$$

$$\beta_x^{(i)} = \sum_{m=1}^{\infty}\sum_{n=1}^{\infty} \overline{\beta}_{xmn}^{(i)}(\xi)\cos(n\varphi)e^{j\omega_{mn}t}$$

$$\beta_\varphi^{(i)} = \sum_{m=1}^{\infty}\sum_{n=1}^{\infty} \overline{\beta}_{\varphi mn}^{(i)}(\xi)\sin(n\varphi)e^{j\omega_{mn}t}$$

$$N_{xx}^{(i)} = h_2 B_{11}^{(2)} \sum_{m=1}^{\infty}\sum_{n=1}^{\infty} \overline{N}_{xxmn}^{(i)}(\xi)\sin\left(\frac{n\pi R_i\varphi}{a_i}\right)e^{i\omega_{mn}t}$$

$$N_{\varphi\varphi}^{(i)} = h_2 B_{11}^{(2)} \sum_{m=1}^{\infty}\sum_{n=1}^{\infty} \overline{N}_{\varphi\varphi mn}^{(i)}(\xi)\sin\left(\frac{n\pi R_i\varphi}{a_i}\right)e^{i\omega_{mn}t}$$

$$N_{x\varphi}^{(i)} = h_2 B_{11}^{(2)} \sum_{m=1}^{\infty}\sum_{n=1}^{\infty} \overline{N}_{x\varphi mn}^{(i)}(\xi)\cos\left(\frac{n\pi R_i\varphi}{a_i}\right)e^{i\omega_{mn}t}$$

$$M_{xx}^{(i)} = h_2^2 B_{11}^{(2)} \sum_{m=1}^{\infty}\sum_{n=1}^{\infty} \overline{M}_{xxmn}^{(i)}(\xi)\sin\left(\frac{n\pi R_i\varphi}{a_i}\right)e^{i\omega_{mn}t}$$

$$M_{\varphi\varphi}^{(i)} = h_2^2 B_{11}^{(2)} \sum_{m=1}^{\infty}\sum_{n=1}^{\infty} \overline{M}_{\varphi\varphi mn}^{(i)}(\xi)\sin\left(\frac{n\pi R_i\varphi}{a_i}\right)e^{i\omega_{mn}t}$$

$$M_{x\varphi}^{(i)} = h_2^2 B_{11}^{(2)} \sum_{m=1}^{\infty}\sum_{n=1}^{\infty} \overline{M}_{x\varphi mn}^{(i)}(\xi)\cos\left(\frac{n\pi R_i\varphi}{a_i}\right)e^{i\omega_{mn}t}$$

$$Q_{xz}^{(i)} = \frac{h_2^2 B_{11}^{(2)}}{R_2} \sum_{m=1}^{\infty}\sum_{n=1}^{\infty} \overline{Q}_{xzmn}^{(i)}(\xi)\sin\left(\frac{n\pi R_i\varphi}{a_i}\right)e^{i\omega_{mn}t}$$

$$Q_{\varphi z}^{(i)} = \frac{h_2^2 B_{11}^{(2)}}{R_2} \sum_{m=1}^{\infty}\sum_{n=1}^{\infty} \overline{Q}_{\varphi zmn}^{(i)}(\xi)\cos\left(\frac{n\pi R_i\varphi}{a_i}\right)e^{i\omega_{mn}t}$$

(i=1,2) **(B.1)**

where i=1,2 indicates the "upper" and "lower" shells, respectively and where the "barred" quantities stand for the dimensionless displacements and the angles of rotation and stress resultants. Also, $\mathbf{j=\sqrt{-1}}$ and the "unbarred" ω_{mn} is the dimensional natural (or circular frequency) of the composite stiffened shell system.

- **<u>Discretized "Dimensionless Fundamental Dependent Variables" and Discretized "Modified Transfer Matrix</u>** $\left[\dot{\mathcal{T}}\right]$ **" (for Part I Region):**

$$\Big\{\left[\overline{U}_1^{(1)}\ \overline{U}_2^{(1)}\ \ .\ .\ \overline{U}_k^{(1)}\right],\left[\overline{V}_1^{(1)}\ \overline{V}_2^{(1)}\ \ .\ .\ \overline{V}_k^{(1)}\right],\ldots,\left[\overline{Q}_{xz1}^{(1)}\ \overline{Q}_{xz2}^{(1)}\ \ .\ .\ \overline{Q}_{xzk}^{(1)}\right];$$
$$\left[\overline{U}_1^{(2)}\ \overline{U}_2^{(2)}\ \ .\ .\ \overline{U}_k^{(2)}\right],\left[\overline{V}_1^{(2)}\ \overline{V}_2^{(2)}\ \ .\ .\ \overline{V}_k^{(2)}\right],\ldots,\left[\overline{Q}_{xz1}^{(2)}\ \overline{Q}_{xz2}^{(2)}\ \ .\ .\ \overline{Q}_{xzk}^{(2)}\right]\Big\}^T$$
$$=\left[\dot{\mathcal{T}}\right]\Big\{\left[\overline{U}_1^{(1)}\ \overline{U}_1^{(1)}\ \ .\ .\ \overline{U}_1^{(1)}\right],\left[\overline{V}_1^{(1)}\ \overline{V}_1^{(1)}\ \ .\ .\ \overline{V}_1^{(1)}\right],\ldots,\left[\overline{Q}_{xz1}^{(1)}\ \overline{Q}_{xz1}^{(1)}\ \ .\ .\ \overline{Q}_{xz1}^{(1)}\right];$$
$$\left[\overline{U}_1^{(2)}\ \overline{U}_1^{(2)}\ \ .\ .\ \overline{U}_1^{(2)}\right],\left[\overline{V}_1^{(2)}\ \overline{V}_1^{(2)}\ \ .\ .\ \overline{V}_1^{(2)}\right],\ldots,\left[\overline{Q}_{xz1}^{(2)}\ \overline{Q}_{xz1}^{(2)}\ \ .\ .\ \overline{Q}_{xz1}^{(2)}\right]\Big\}^T$$

(B.2)

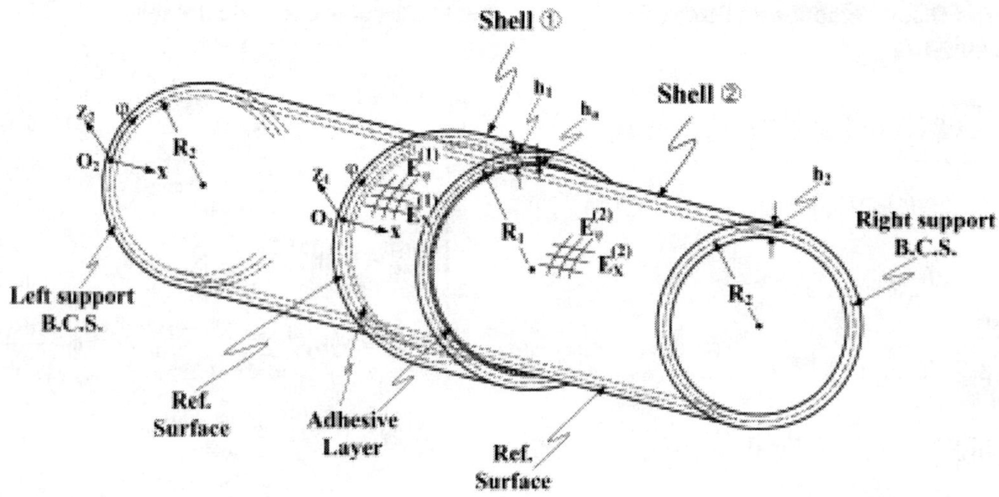

(a)

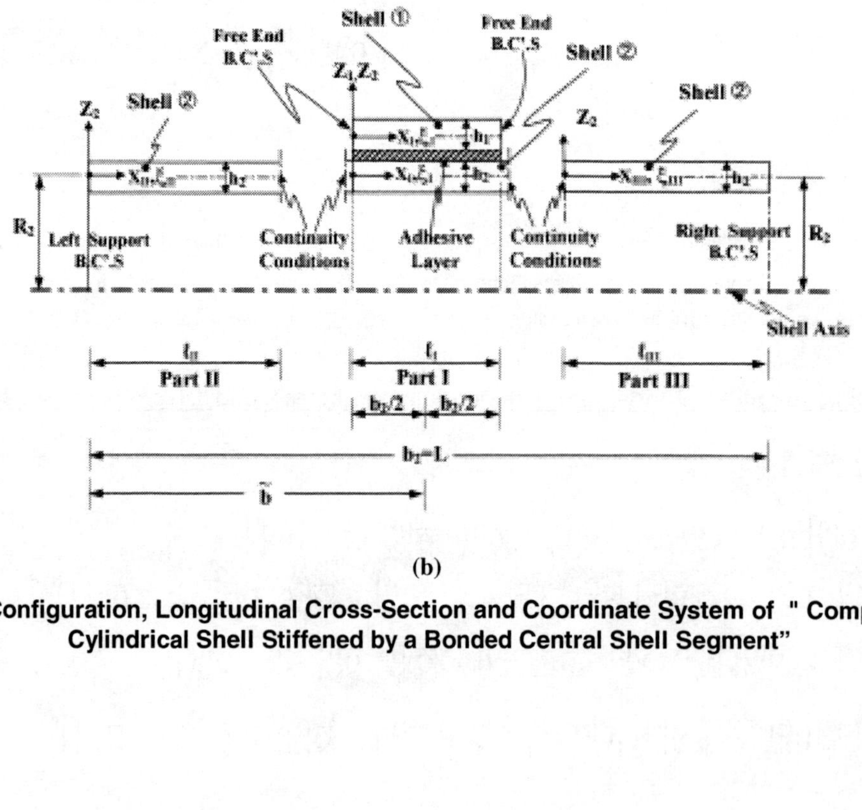

(b)

Figure 1. General Configuration, Longitudinal Cross-Section and Coordinate System of " Composite Full Circular Cylindrical Shell Stiffened by a Bonded Central Shell Segment"

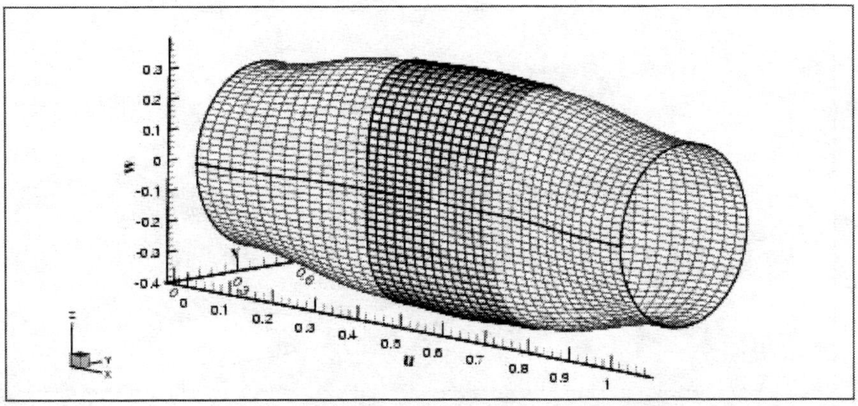

a) First Mode with $\overline{\Omega}_1 = \overline{\omega}_{12} = 14.959$

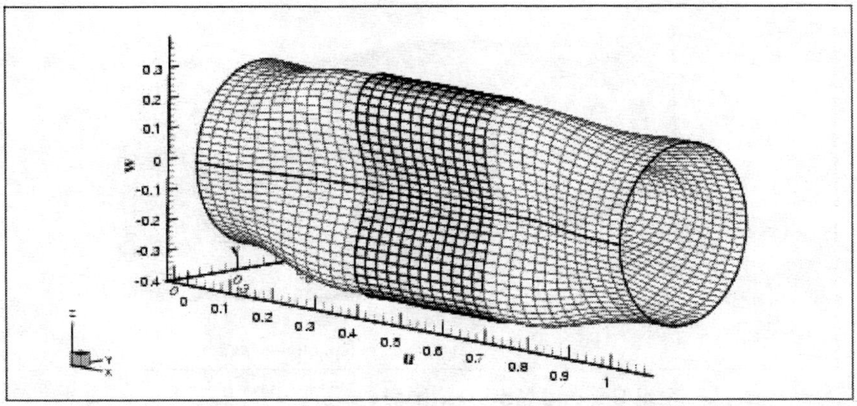

b) Second Mode with $\overline{\Omega}_2 = \overline{\omega}_{13} = 24.573$

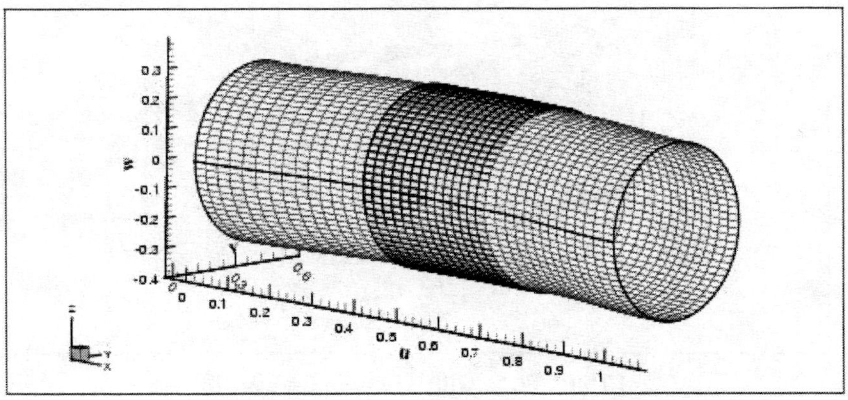

c) Third Mode with $\overline{\Omega}_3 = \overline{\omega}_{11} = 36.759$

Figure 2. Mode Shapes and Natural Frequencies of " Composite Full Circular Cylindrical Shell Stiffened by a Bonded Central Shell Segment"
(Upper Shell= Shell 1= Kevlar-Epoxy, Base or Lower Shell=Shell 2= Graphite-Epoxy)
((FFCC) B.C.'s in y-direction, "Hard" Adhesive, Joint Length= ℓ_I = 0.30 m, R=0.30 m, L=1.0 m)

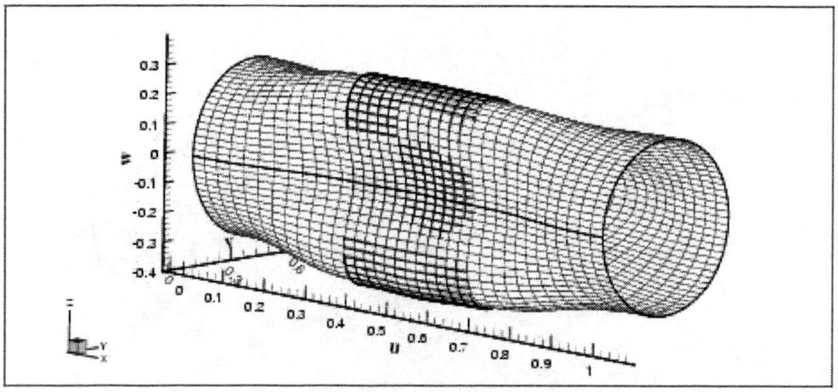

a) First Mode with $\overline{\Omega}_1 = \overline{\omega}_{13} = 12.012$

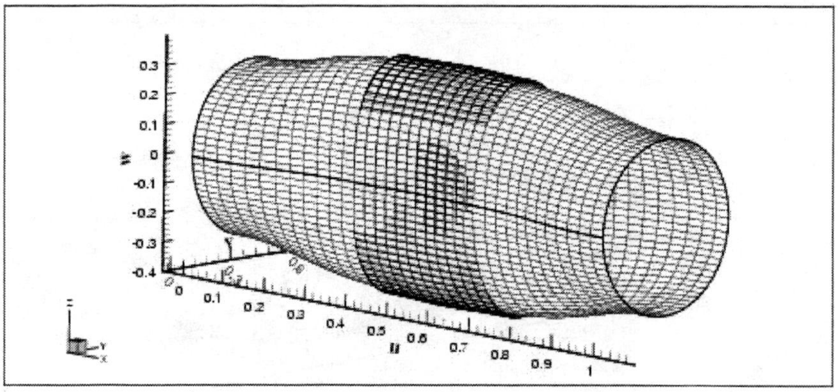

b) Second Mode with $\overline{\Omega}_2 = \overline{\omega}_{12} = 12.335$

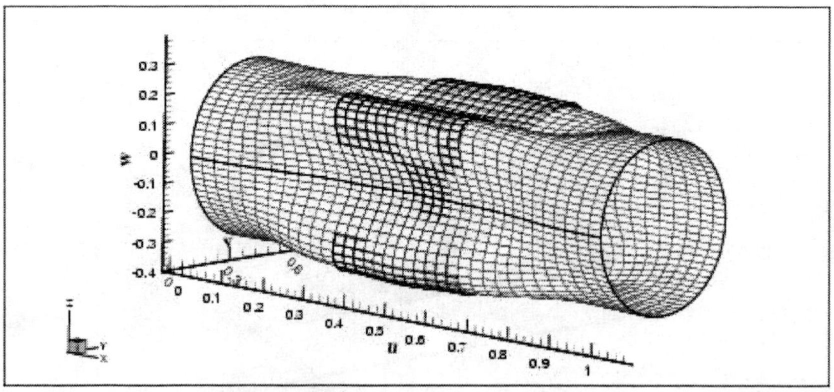

c) Third Mode with $\overline{\Omega}_3 = \overline{\omega}_{14} = 29.199$

Figure 3. Mode Shapes and Natural Frequencies of " Composite Full Circular Cylindrical Shell Stiffened by a Bonded Central Shell Segment"
(Upper Shell= Shell 1= Kevlar-Epoxy, Base or Lower Shell=Shell 2= Graphite-Epoxy)
((FFCC) B.C.'s in y-direction, "Soft" Adhesive, Joint Length= ℓ_I = 0.30 m, R=0.30 m, L=1.0 m)

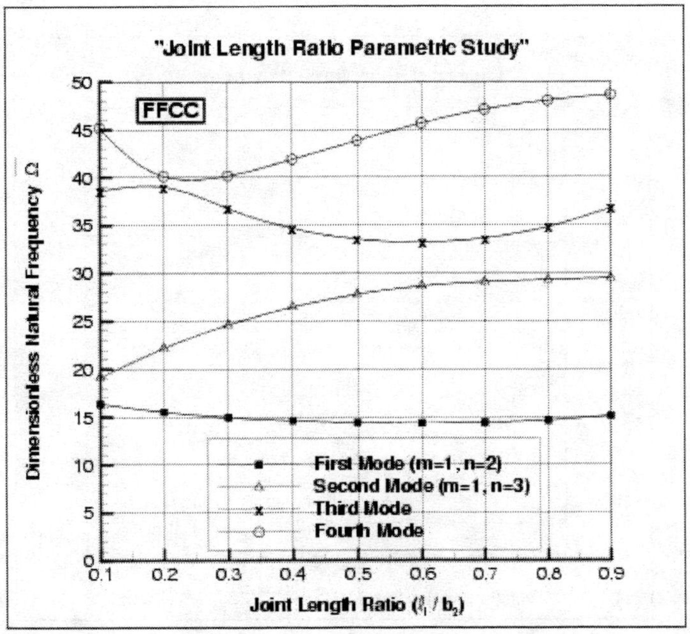

Figure 4. Natural Freq. versus "Joint Length Ratio ℓ_I/L" in " Composite Full Circular Cylindrical Shell Stiffened by a Bonded Central Shell Segment"
(Shell 1= Kevlar-Epoxy, Shell 2= Graphite-Epoxy)
("Hard" Adhesive, L=1.0 m., R=0.3 m., ℓ_I varies)
(FFCC Boundary Conditions in x-direction)

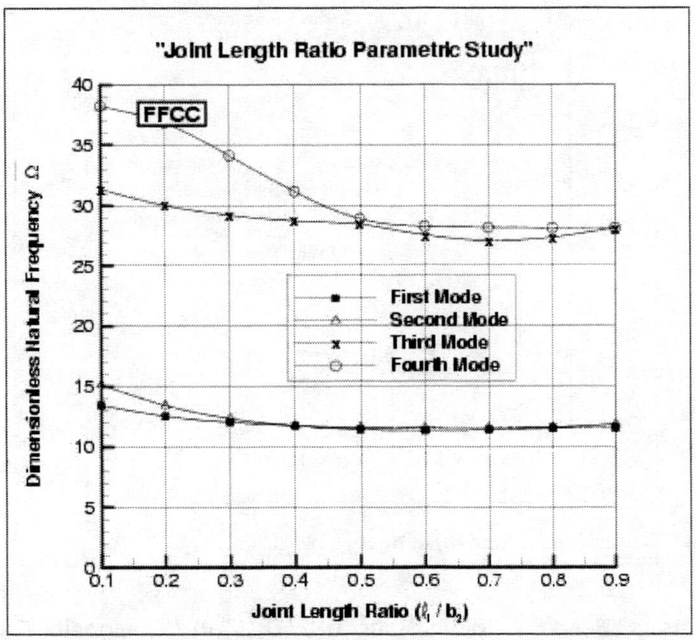

Figure 5. Natural Freq. versus "Joint Length Ratio ℓ_I/L" in " Composite Full Circular Cylindrical Shell Stiffened by a Bonded Central Shell Segment"
(Shell 1= Kevlar-Epoxy, Shell 2= Graphite-Epoxy)
("Soft" Adhesive, L=1.0 m., R=0.3 m., ℓ_I varies)
(FFCC Boundary Conditions in x-direction)

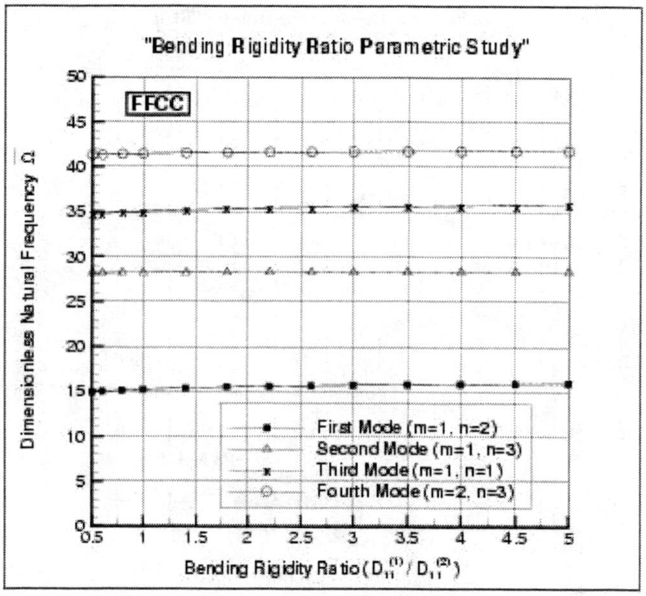

Figure 6. Natural Freq. versus "Bending Rigidity Ratio $D_{11}^{(1)}/D_{11}^{(2)}$" in " Composite Full Circular Cylindrical Shell Stiffened by a Bonded Central Shell Segment"
("Hard" Adhesive, Shell 1= Stiffness Varies, Shell 2= Graphite-Epoxy)
(L=1.0 m., R=0.3 m., ℓ_I =0.3 m.)
(FFCC Boundary Conditions in x-direction)

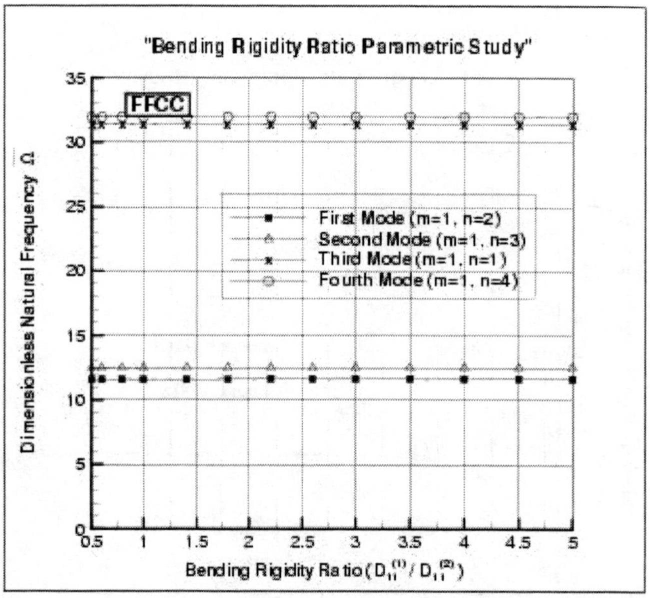

Figure 7. Natural Freq. versus "Bending Rigidity Ratio $D_{11}^{(1)}/D_{11}^{(2)}$" in " Composite Full Circular Cylindrical Shell Stiffened by a Bonded Central Shell Segment"
("Soft" Adhesive, Shell 1= Stiffness Varies, Shell 2= Graphite-Epoxy)
(L=1.0 m., R=0.3 m., ℓ_I =0.3 m.)
(FFCC Boundary Conditions in x-direction)

Proceedings of IMECE04
2004 ASME International Mechanical Engineering Congress and Exposition
November 13-20, 2004, Anaheim, California USA

IMECE2004-59883

PERFORMANCE EVALUATION OF NOTCHED BIAXIAL BRAIDED COMPOSITES

Jitendra S. Tate, Ajit D. Kelkar, and Ronnie Bolick
Department of Mechanical Engineering
North Carolina A & T State University
626 McNair Hall, 1601 E Market Street
Greensboro, North Carolina, 27411, USA
Phone: 336-334-7620 x324, Fax: 336-334-7417, Email: kelkar@ncat.edu

ABSTRACT

Braided composites have good properties in mutually orthogonal directions, more balanced properties than traditional tape laminates, and have potentially better fatigue and impact resistance due to the interlacing. Another benefit is reduced manufacturing cost by reducing part count. Because of these potential benefits braided composites are being considered for various applications ranging from primary/secondary structures for aerospace structures [1]. These material systems are gaining popularity, in particular for the small business jets, where FAA requires take off weights of 12,500 lb. or less. The new process, Vacuum Assisted Resin Transfer Molding (VARTM), is low cost, affordable and suitable for high volume manufacturing environment. Recently the aircraft industry has been successful in manufacturing wing flaps, using carbon fiber braids and epoxy resin and the VARTM process. To utilize these VARTM manufactured braided materials to the fullest advantage (and hence to avoid underutilization), it is necessary to understand their behavior under different loading and environmental conditions. This will reduce uncertainty and hence reduce the factor of safety in the design. It is well known fact that the strength of the composite structure reduces because of discontinuities and abrupt changes in the cross-section. Accurate knowledge of strength and failure mechanism of notched and unnotched composites is very important for design of composite structures. This research addresses the behavior of notched braided composites under static tensile loading.

KEY WORDS

VARTM, Braid, Notched

INTRODUCTION

Braided composites have better interlaminar shear strength. It has been observed that, they do not exhibit delamination under static tensile and tension-tension fatigue loading [2]. Secondly, the mechanical properties of braided composites are highly sensitive to the braid angle. Therefore the failure mechanism of notched braided composites might be different than that of notched laminates or woven composites.

The present research addresses the behavior of notched braided composites under static tensile loading. In the present investigation, all the braided composite panels were manufactured using low cost vacuum assisted resin transfer molding (VARTM) process.

OBJECTIVES

(a) To fabricate 2 x 2 biaxial braided composites using the VARTM low cost manufacturing process
(b) To perform static tension tests on notched and unnotched specimens
(c) To study the failure mechanism of notched braided composites

BIAXIAL BRAIDS

Braided fiber architecture resembles a hybrid of filament winding and weaving. Like filament winding, tubular braid features seamless fiber continuity from end to end of a part. Like woven materials, braided fibers are mechanically interlocked with one another. The combination, however, is quite extraordinary. When functioning as a composite reinforcement, braid exhibits remarkable properties because it is highly efficient in distributing loads. Because all the fibers within a braided structure are continuous and mechanically locked, braid has a natural mechanism that evenly distributes load throughout the structure. Braided structures are good in fatigue and impact. They also exhibit improved interlaminar shear properties [1]. Braids are available in different forms such as sleevings or tubes, flat braids or tapes. Braids are also available in different architecture like biaxial and triaxial.

The biaxial tubes form of the braid is the most popular form and chosen by composite manufacturers since it allows for predictable, consistent lay-up and conforms to any shape. Depending upon the fiber orientation termed as 'braid angle' (Refer Figure 1), the thickness and the yield of the braid vary in the panel. The variation in the properties like thickness and areal weight according to fiber orientation is predictable and repeatable [1]. The biaxial braid offers natural conformability. This quality makes braid the ideal preform since it takes on the exact shape of the part that it is reinforcing. Cutting, stitching, or manipulation of fiber placement is not needed, as in the use of woven reinforcing. Another benefit is reduced part count. Therefore, braids are cost competitive once overall cost analysis is performed [1]. With regard to strength and stiffness, properly molded biaxial braided composites exhibit properties that are comparable to unidirectional tape laminates of the same orientation and fiber volume [3].

VARTM FABRICATION PROCESS

VARTM is a single/double sided tooling process where the dry preform is placed into the tool and vacuum bagged in conjunction with resin distribution and vacuum distribution lines. A low viscosity resin is drawn into the preform through the aid of vacuum. In VARTM process, the flow of resin occurs in plane as well as in the transverse directions to perform [Refer Figure 2]. The permeability of the preform, fiber architecture and fabric crimp has an influence on the wetting of the fabric. This process is being currently used in many of the applications in the general aviation industry, defense sector and in the transport industry.

This process has certain advantages as listed below:

- Simple low cost single sided tooling
- Very large and complex parts are practical
- High fiber volume fraction than hand lay-up
- Reduced environmental concerns than hand lay-up

Recently the aircraft industry has been successful in manufacturing wing flaps, using carbon fiber braids and epoxy resin and the VARTM process. This process is also studied for complete automation, sensing and control for manufacturing large-scale composite parts [4]. This process has been proved cost effective in manufacturing of ballistic, multifunctional, composite material systems and structures [5]. It is decided to use this method for the present research work.

MATERIAL SYSTEM

Biaxial carbon (AS4) slit sleeves with braid angle of 30° manufactured by A & P Technology, Inc. and EPON 9504 epoxy resin manufactured by Resolution Performance Products, Inc. were used to fabricate braided composites. There were four plies. The total thickness of braided composite was 3.56 mm (0.14 in.) with overall fiber volume fraction of 0.53+/-0.05.

STATIC TENSION TESTS (ASTM D3039)

These tests were performed as per specifications mentioned in the ASTM D3039. The specimen size was 31.75 mm (1.25 in.) wide and 254 mm (10 in.) long. The holes were drilled using special CBN diamond core drill bits to avoid delaminations. Most of the bolted structural applications involve use of quarter inch diameter holes, hence in the present investigation two hole diameters: 4.76 mm (0.1875 in.) and 6.35 mm (0.25 in.) were used. All the tests were performed on MTS testing machine with Instron controller 8500 and Instron hydraulic grips. Displacement control static tests were performed to evaluate ultimate tensile strength. The displacement rate used was 1.27 mm/min (0.05 in/min) as recommended by ASTM.

RESULTS AND DISCUSSIONS

The results of the static tensile tests are presented in the Table 1. The average tensile strength of five unnotched specimens was 64.09 ksi (441.84 MPa) [2]. Two notched specimens were tested in each category.

Table 1 Static Test Results

Hole Size (d), mm (in.)	(d/W), Ratio	Average Net-section Strength, ksi (MPa)
0	0.00	64.09 (441.84)
4.76 (0.1875)	0.15	61.86 (426.46)
6.35 (0.25)	0.20	60.31 (415.77)

It should be noted that the braid angle variation within a specimen was 30+/-1° and overall fiber volume fraction between the specimens was 0.53+/-0.05. Thus the specimens exhibited considerable heterogeneity. It is possible to establish relation for ultimate tensile strength as a function of braid angle and fiber volume fraction [6]. For accurate quantitative understanding of stress concentration factors it is essential to establish such relation when the sample size is small as in this research. The purpose of present research is to understand failure mechanism and not to quantify stress concentration factors.

The failure of all the specimens, notched and unnotched did not exhibit edge delamination and/or matrix cracking and occurred catastrophically. In case of notched specimens, at certain load the hole became elliptical with major axis aligned with loading direction. The cracks initiated near the holes (around 90° with respect to load direction) and there onward failure propagated in the transverse direction. Figure 3 displays the progressive failure of the notched specimen.

Earlier investigators have reported stress concentration values of 2.6 and 2.4 for d/W ratio of 0.15 to 0.20 respectively for isotropic materials [7]. On the other hand these values are as high as 9.0 for the unidirectional laminates and 6.0 for the angle ply laminates under uniaxial tensile loadings [8]. Since braid resembles angle ply stacked laminates, it was expected that the braided composites with hole would exhibit much lower failure strength than obtained in the present investigation. Furthermore, the failure pattern of notched and unnotched braided composite was similar (Refer Figure 4). This may be due to the high interlminar shear strength of the braided composites.

CONCLUSIONS

(1) The 2 x 2 biaxial braided composites were successfully fabricated using low cost VARTM process.

(2) The notched and unnotched braided composite specimens failed without showing any edge delamination and/or noticeable matrix cracking.

(3) It is recommended that, the failure of notched braided composites should be observed using high-speed camera under static tensile and fatigue loading.

ACKNOWLEDGEMENT

Federal Aviation Administration sponsored the work reported herein under a contract DTFA03-01-C00033. This financial help is gratefully acknowledged.

REFERENCES

[1] http://www.braider.com web site of A & P Technology, Inc.

[2] Tate, Jitendra S., 2004, "Performance Evaluation and Modeling of Braided Composites", Ph.D. thesis, NC A & T State University, Greensboro, NC

[3] Falzon, Paul J., and Herszberg, Israel, 1998, "Mechanical performance of 2-D braided carbon/epoxy composites", Composites Science and Technology, v 58, n 2, p 253-265

[4] Heider, Dirk, Hofmann, C., and Gillespie, J.W. Jr., 2000, "Automation and control of large-scale composite parts by VARTM processing", 45th International SAMPE Symposium and Exhibition, CA, USA, p 1567-1575

[5] Rigas, E.J., Walsh, S. M., Spurgeon, W.A., 2001, "Development of a novel processing technique for Vacuum Assisted Resin Transfer Molding (VARTM)", 46th International SAMPE Symposium and Exhibition 2001 a Materials and Processes Odyssey, CA, p 1086-1094

[6] Tate J.S., Kelkar A.D., and Kelkar V. A., 2004, " Failure Analysis of Biaxial Braided Composites under Fatigue Loading", The 15th European Conference of Fracture, Stockholm, Sweden

[7] Smith, Clarence R., 1963, "Tips on Fatigue," Report NAVWEPS 00-25-559, Bureau of Naval Weapons, Washington, D.C.

[8] Greszczuk, L. B., 1972, "Stress Concentrations and Fatigue Criteria for Orthotropic Plates with Circular Openings," 2nd Conference on Composite Materials: Testing and Design, ASTM STP 497, pp 363-381

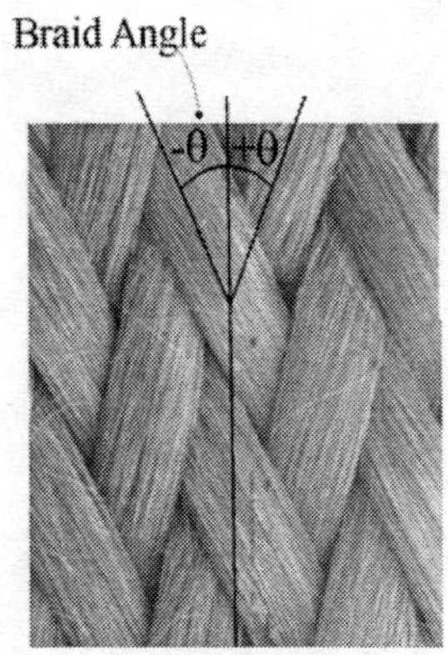

Figure 1. Biaxial braid

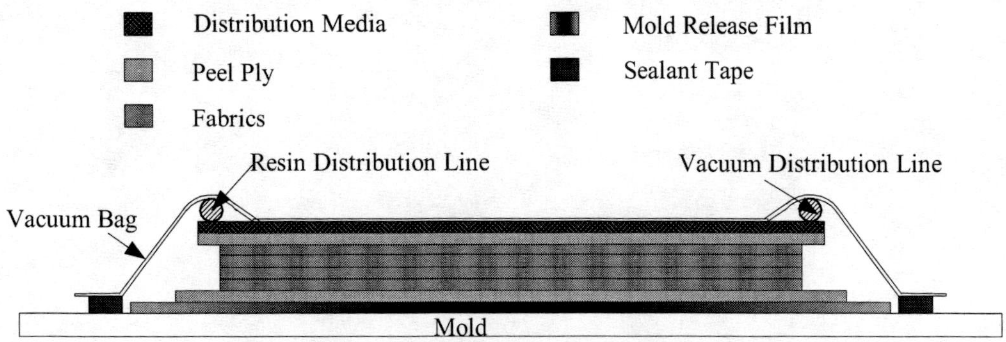

Figure 2. Schematic of VARTM

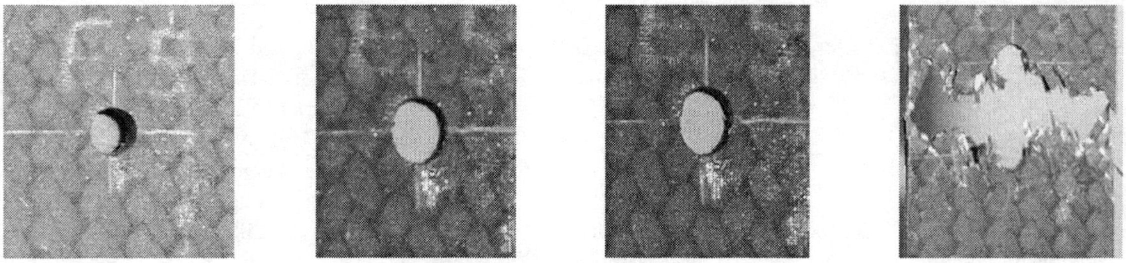

(a) (b) (c) (d)

Figure 3. Sequence of failure of notched braided composite (a) Hole before loading (b) Elongated hole in the direction of loading (c) Crack initiation near the hole (d) Breakage along transverse direction

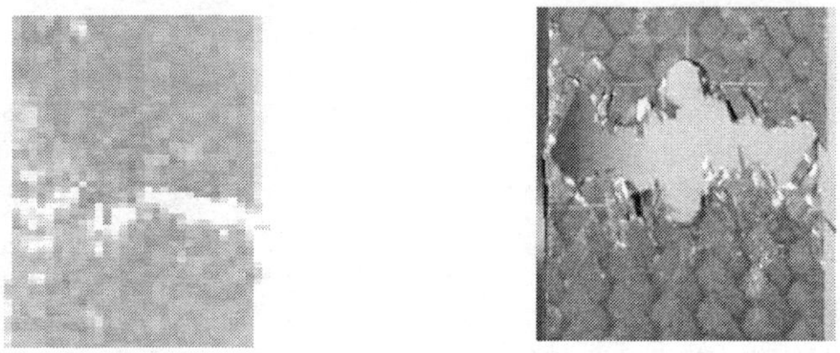

(a) (b)

Figure 4. (a) Failure of unnotched specimen (b) Failure of notched specimen

Proceedings of IMECE04
2004 ASME International Mechanical Engineering Congress and Exposition
November 13-20, 2004, Anaheim, California USA

IMECE2004-60014

PROCEDURES FOR THE VERIFICATION AND VALIDATION OF WORKING MODELS FOR STRUCTURAL SHELLS

Barna A. Szabó*
Center for Computational Mechanics
Department of Mechanical and Aeronautical Engineering
Washington University
St. Louis, Missouri 63130
Email: szabo@me.wustl.edu

Daniel E. Muntges
Center for Computational Mechanics
Department of Mechanical and Aeronautical Engineering
Washington University
St. Louis, Missouri 63130
Email: dem2@cec.wustl.edu

ABSTRACT

Verification of computed data and validation of mathematical models for structural shells with reference to particular applications and goals of computation are discussed and illustrated through the example of a cylinder-to-cylinder intersection problem investigated experimentally and analytically at Oak Ridge National Laboratory.

1 Introduction

In structural shells, such as pressure vessels and piping systems, there are regions where the kinematic assumptions incorporated in classical shell models do not hold. These are the neighborhoods of nozzles and shell intersections, support attachments, stiffeners, cut-outs, and regions where the curvature abruptly changes. From the point of view of engineering design and analysis, these are the typical regions of primary interest, however.

In formulating mathematical models, shell-like structures should be viewed as fully three-dimensional solid bodies that allow the imposition of restrictions on the transverse variation of displacement vector components in certain regions. The imposition of such restrictions reduces a three-dimensional problem to a two-dimensional one in the case of plates and shells; to a one-dimensional problem in the case of beams and arches. An important practical problem is to identify the proper formulation for a particular application, that is, a formulation that simplifies the problem by dimensional reduction without affecting the data of interest significantly.

In our terminology a working model is a mathematical model of some physical system or process comprised of a formulation and data pertaining to geometric description, material properties, loading and constraints. Associated with a working model is an exact solution.

Common to all mathematical models based on the principles of continuum mechanics are the conservation of momentum (in static problems the equations of equilibrium), the strain-displacement relations and constitutive laws. A particular application involves selection of a working model.

Any working model should be viewed as a special case of a general model, called state of the art model. By definition, a state of the art model accounts for all physical laws that pertain to the system or process being modeled. Selection of a particular working model involves acceptance of certain limitations that may or may not be justified, given the goals of computation. In general it is not known a priori whether the limitations of a particular working model are justified. Therefore it is necessary to perform tests a posteriori to determine whether a particular working model is suitable with respect to the goals of computation. If the tests show that a working model is inadequate then a more comprehensive working model must be chosen and the process repeated.

In principle, one could continue this process indefinitely until all relevant attributes of the physical system or process being modeled are taken into account. In reality however the more

*The Albert P. and Blanche Y. Greensfelder Professor of Mechanics.

comprehensive a working model, the more descriptive data that characterize the geometric description, material properties and boundary conditions are needed. Often the needed parameters are not available, or would be difficult to determine. The cognitive as well as the stochastic uncertainties increase with the complexity of the working model. Therefore a highly detailed working model is generally not feasible. Consequently is some degree of uncertainty associated with any attempt to model a physical system or process by mathematical means. On the other hand, much useful information can be developed through systematic evaluation of sequences of idealized working models with reference to the data of interest.

Errors associated with the choice of model are called errors of idealization whereas errors associated with finite element or other numerical approximation of the exact solution of the model are called errors of discretization. The process by which errors of idealization are controlled is called validation. The process by which errors of discretization are controlled is called verification. In this paper control of both the errors of idealization and discretization are discussed with reference to a model problem for which detailed experimental information of high quality is available.

The model problem discussed in this paper is the first of four models investigated by experimental and analytical means at Oak Ridge National Laboratory in the 1970's [1], [2]. The test articles consisted of two thin-walled circular cylindrical shells intersecting at right angles. The cylindrical shells were made of carbon steel. The test articles are idealized representations of shell intersections that occur in piping and pressure vessel systems in the sense that there were no transitions, fillets or reinforcements at the junctions. The stated objective of the experiments was to obtain stress data for comparison with the predictions of thin shell theory.

A hierarchic family of working models that allow virtually arbitrary representation of the transverse variation of the displacement vector components is described in the following. The lowest member of the hierarchy is similar to the Naghdi shell model. The highest member is the fully three-dimensional representation. Selection of a working model for a particular application is based on feedback information that includes the results of tests conducted with the objective to determine whether the data of interest are dependent on the restrictive assumptions incorporated in the model.

2 Comparison between the results of experiments and finite element analyses

We are concerned only with working models that have unique exact solutions. We denote the exact solution of working model m by $\vec{u}_m$. We are interested in some set of measurable data, such as displacements and strains, in specific points. We will denote these data corresponding to $\vec{u}_m$ by $\Phi_i(\vec{u}_m)$, $i = 1,2,\ldots$. Of course, the exact solution is generally unknown, only an approximation to the exact solution, computed by the finite element method or some other numerical method, is known. We denote the approximate solution by $\vec{u}_a$ and the data computed from $\vec{u}_a$ by $\Phi_i(\vec{u}_a)$ $(i = 1,2,\ldots)$. The error of numerical approximation is denoted by e_i^{num}:

$$e_i^{\text{num}} := \Phi_i(\vec{u}_m) - \Phi_i(\vec{u}_a) \tag{1}$$

We denote the experimental measurement corresponding to $\Phi_i(\vec{u}_m)$ by $\Phi_i^\star$. Associated with $\Phi_i^\star$ is a random and, possibly, systematic error in the experimental observation which is denoted by e_i^{exp}. We denote the ideal (error-free) measurement by $\bar{\Phi}_i$. Therefore

$$e_i^{\text{exp}} := \bar{\Phi}_i - \Phi_i^\star. \tag{2}$$

Let us assume that all necessary information concerning the object of an experiment is available. Then, in comparing experimental data with predictions based on a working model, three sources of error are present: (a) errors in the working model itself, including the formulation, variations in dimensions, material properties and boundary conditions; (b) systematic and random errors in the experimental observations e_i^{exp}, and (c) errors in the numerical approximation of the solution of the mathematical model e_i^{num}, called the errors of discretization.

We will consider mathematical model m a reasonable representation of the physical reality of an experiment if

$$|\Phi_i(\vec{u}_m) - \bar{\Phi}_i| \le \tau_i |\Phi_i(\vec{u}_m)|, \qquad i = 1,2,\ldots \tag{3}$$

where τ_i are prescribed tolerances. Using equations (1) and (2) and the triangle inequality we have

$$\begin{aligned} |\Phi_i(\vec{u}_m) - \bar{\Phi}_i| =& |\Phi_i(\vec{u}_a) + e_i^{\text{num}} - \Phi_i^\star - e_i^{\text{exp}}| \\ \le& |\Phi_i(\vec{u}_a) - \Phi_i^\star| + |e_i^{\text{num}}| + |e_i^{\text{exp}}|. \end{aligned} \tag{4}$$

It is seen that replacement of $\Phi_i(\vec{u}_m)$ with $\Phi_i(\vec{u}_a)$ and $\bar{\Phi}_i$ with $\Phi_i^\star$ is permissible only if it can be shown that the relative errors $|e_i^{\text{exp}}/\Phi_i(\vec{u}_m)|$ and $|e_i^{\text{num}}/\Phi_i(\vec{u}_m)|$ are much smaller than τ_i. If this condition is not satisfied then it is possible that large errors in the mathematical model are obscured by similarly large errors in the numerical approximation and/or the experimental observations. Therefore estimation and control of e_i^{num} and e_i^{exp} are essential.

Remark 2.1. We have assumed that all information necessary for the formulation of a working model is available. This requirement can be difficult to meet. Requirements for validation experiments are discussed in [3].

Remark 2.2. Successful prediction of the outcomes of one or more physical experiments does not prove that a working model will provide correct predictions for other experiments. On the other hand, failure to correctly predict the outcome of any one experiment is sufficient evidence for rejection. Therefore the grounds for rejection of working models are well established (in principle) but it is not possible to prove through experimentation that a particular working model is correct. With each successful prediction of the outcome of a physical experiment the probability that the working model will give correct results increases but guarantees cannot be given. In other words, acceptance of a working model is tentative, whereas rejection is conclusive.

3 Formulation

The generalized formulation of mathematical models for shell-like structures is discussed from the point of view of algorithmic structure and computer implementation.

The development of classical shell models was strongly influenced by the limitations of the methods available for solving the resulting equations. The use of curvilinear coordinates allowed the treatment of shells with simple geometric description, such as cylindrical, spherical and conical shells, by classical methods, subject to the assumption that the thickness of the shell is small in relation to its other dimensions. Such limitations no longer exist. Currently the main concern is the selection of a working model suitable for a particular application, given the goals of computation. In general, it is not known a priori, whether a particular working model is suitable for the purposes of an analysis. Therefore there has to be an a posteriori evaluation of the data of interest and, if necessary, means must be available to modify the working model

Adaptive methods continue to be a focus of interest in finite element analysis. The term adaptivity has various meanings: In the broad (engineering) sense adaptivity is understood to be a process devised for efficient control of the errors of idealization and the errors of discretization to guarantee that the data of interest are computed to within a specific tolerance. In this paper we address adaptivity from this perspective, the objective being to control both the errors of discretization and the errors of idealization. This requires a hierarchic framework of models, i.e., models of increasing complexity, allowing systematic removal of limitations guided by feedback information.

3.1 Notation

We denote the unit vectors of the Cartesian reference system by $\vec{e}_x$, $\vec{e}_y$, $\vec{e}_z$ and the coordinates of an arbitrary point in $\mathbb{R}^3$ by $\vec{r} = x\vec{e}_x + y\vec{e}_y + z\vec{e}_z$. The material points of a solid body are denoted by Ω, its boundary points by $\partial\Omega$ and the differential surface element by dS. The displacement vector of a point denoted by

$$\vec{u}(x,y,z) := u_x(x,y,z)\vec{e}_x + u_y(x,y,z)\vec{e}_y + u_z(x,y,z)\vec{e}_z. \quad (5)$$

The vector of strain tensor components corresponding to $\vec{u}$ is denoted by $\{\varepsilon\}$. The relationship between $\vec{u}$ and $\{\varepsilon\}$ is given by $\{\varepsilon\} = [D]\vec{u}$ where $[D]$ the differential operator associated with small strain elasticity. The vector of stress tensor components is denoted by $\{\sigma\}$. The relationship between $\{\sigma\}$ and $\{\varepsilon\}$ is given by Hooke's law: $\{\sigma\} = [E]\{\varepsilon\}$ where $[E]$ is the elastic material stiffness matrix. Traction vectors acting on $\partial\Omega$ are denoted by $\vec{T}$. The virtual work of internal stresses is denoted by $B(\vec{u},\vec{v})$, defined as follows:

$$B(\vec{u},\vec{v}) := \int_\Omega ([D]\vec{v})^T [E][D]\vec{u}\, d\Omega \quad (6)$$

and the virtual work of external forces is denoted by

$$F(\vec{v}) := \int_\Omega \vec{F}\cdot\vec{v}\, d\Omega + \int_{\partial\Omega} \vec{T}\cdot\vec{v}\, dS + \int_\Omega ([D]\vec{v})^T [E]\{c\}\tau\, d\Omega \quad (7)$$

where $\{c_t\} := \{c_t\ c_t\ c_t\ 0\ 0\ 0\}^T$, c_t is the coefficient of thermal expansion and $\tau = \tau(x,y,z)$ is the temperature.

The energy space is defined by

$$E(\Omega) := \{\vec{u}\,|\,B(\vec{u},\vec{u}) \le C < \infty\} \quad (8)$$

where C is some positive constant. The energy norm is defined by

$$\|\vec{u}\|_{E(\Omega)} := \left(\frac{1}{2}B(\vec{u},\vec{u})\right)^{1/2}. \quad (9)$$

The generic form of the principle of virtual work is stated as follows: Find $\vec{u} \in E(\Omega)$ such that

$$B(\vec{u},\vec{v}) = F(\vec{v}) \qquad \text{for all } \vec{v} \in E(\Omega). \quad (10)$$

Specific statements of the principle of virtual work depend on the boundary conditions. For details we refer to [4]. In the generic case, and whenever the prescribed displacement constraints are insufficient to prevent all possible rigid body displacements, the solution of Eqn. (10) is unique up to the missing rigid body displacements.

The mid-surface of a shell is assumed to be a smooth function in $\mathbb{R}^3$, given in terms of two parameters α, β:

$$\vec{\rho} = x(\alpha,\beta)\vec{e}_x + y(\alpha,\beta)\vec{e}_y + z(\alpha,\beta)\vec{e}_z, \qquad (\alpha,\beta) \in \omega \subset \mathbb{R}^2 \quad (11)$$

where ω is a connected domain with piecewise smooth boundary. The curvilinear basis vectors are

$$\vec{e}_\alpha := \frac{\partial\vec{\rho}}{\partial\alpha}\Big/\left|\frac{\partial\vec{\rho}}{\partial\alpha}\right|, \qquad \vec{e}_\beta := \frac{\partial\vec{\rho}}{\partial\beta}\Big/\left|\frac{\partial\vec{\rho}}{\partial\beta}\right|, \\ \vec{e}_n := \frac{\partial\vec{\rho}}{\partial\alpha}\times\frac{\partial\vec{\rho}}{\partial\beta}\Big/\left|\frac{\partial\vec{\rho}}{\partial\alpha}\times\frac{\partial\vec{\rho}}{\partial\beta}\right|. \tag{12}$$

The vector $\vec{e}_n$ is the unit normal. The vectors $\vec{e}_\alpha$, $\vec{e}_\beta$ are the normalized covariant basis vectors of the tangent plane. These basis vectors are generally not orthogonal, however in some important special cases, such as cylindrical surfaces given in cylindrical coordinates, they are. The triplet $(\vec{e}_\alpha\ \vec{e}_\beta\ \vec{e}_n)$ defines the covariant basis.

The material points of the shell are the set of points Ω, defined as follows:

$$\Omega := \{\vec{r}\,|\,\vec{r} = \vec{\rho}(\alpha,\beta) + \nu\vec{e}_n,\ (\alpha,\beta)\in\omega,\ -t/2 < \nu < t/2\} \tag{13}$$

where t is the thickness of the shell.

The displacement vector components are given either in terms of the covariant basis vectors:

$$\vec{u}_{(\alpha)} := u_\alpha(\alpha,\beta,\nu)\vec{e}_\alpha + u_\beta(\alpha,\beta,\nu)\vec{e}_\beta + u_n(\alpha,\beta,\nu)\vec{e}_n \tag{14}$$

or the Cartesian basis vectors:

$$\vec{u}_{(x)} := u_x(\alpha,\beta,\nu)\vec{e}_x + u_y(\alpha,\beta,\nu)\vec{e}_y + u_z(\alpha,\beta,\nu)\vec{e}_z. \tag{15}$$

The displacement components given in the covariant basis can be transformed to the (global) Cartesian basis by the matrix operation:

$$\vec{u}_{(x)} = [R]\vec{u}_{(\alpha)} \tag{16}$$

where the columns of $[R]$ are the unit vectors $\vec{e}_\alpha$, $\vec{e}_\beta$ and $\vec{e}_n$.

3.2 Hierarchic shell models

For hierarchic shell models the displacement vector components are given in terms of the covariant basis in the following form:

$$u_\alpha := \sum_{i=0}^{m_\alpha} u_{\alpha|i}(\alpha,\beta)\phi_i(\nu) \\ u_\beta := \sum_{i=0}^{m_\beta} u_{\beta|i}(\alpha,\beta)\phi_i(\nu) \\ u_n := \sum_{i=0}^{m_n} u_{n|i}(\alpha,\beta)\phi_i(\nu) \tag{17}$$

where $\phi_i(\nu)$ are called director functions. When the material is isotropic then $\phi_i(\nu)$ are polynomials; when the shell is laminated then $\phi_i(\nu)$ are piecewise polynomials [5], [6], [4], [7], [8]. Eq. (17) represents a semi-discretization of the problem of fully three-dimensional elasticity, in that $\phi_i(\nu)$ are fixed and thus the problem is reduced from a three-dimensional problem to a two-dimensional one. A particular model is characterized by the indices (m_α, m_β, m_n).

In the classical treatment of shells the curvilinear basis vectors are retained throughout the analysis. In numerical treatment by the finite element method a generalized formulation is used, most commonly the principle of virtual work. The algorithmic structure becomes simpler if the displacement components given in the covariant basis are transformed to the (global) Cartesian basis by Eqn. (16).

The generic form of the virtual work of internal stresses is given by the expression

$$B(\vec{u}_{(\alpha)},\vec{v}_{(\alpha)}) := \int_\omega\int_{-t/2}^{+t/2} \left([\tilde{D}][R]\vec{v}_{(\alpha)}\right)^T [E][\tilde{D}][R]\vec{u}_{(\alpha)}\,d\nu\,d\omega \tag{18}$$

where $[\tilde{D}]$ is the differential operator that transforms the displacement vector components given in terms of the curvilinear coordinates $(\alpha,\ \beta,\ \nu)$ to the Cartesian strain tensor components: $\{\varepsilon\} = [\tilde{D}]\vec{u}$. It is assumed that the material is isotropic. If the material is not isotropic and its reference frame differs from the global Cartesian coordinate system, then $[E]$ must be transformed into the global Cartesian coordinate system.

The generic form of the virtual work of external forces is:

$$F(\vec{v}_{(\alpha)}) := \int_\omega\int_{-t/2}^{+t/2} \left([R]\vec{F}_{(\alpha)}\right)^T [R]\vec{v}_{(\alpha)}\,d\nu\,d\omega \\ + \int_{\partial\Omega} \left([R]\vec{T}_{(\alpha)}\right)^T [R]\vec{v}_{(\alpha)}\,dS \\ + \int_\omega\int_{-t/2}^{+t/2} \left([\tilde{D}][R]\vec{v}_{(\alpha)}\right)^T [E]\{c\}\tau\,d\nu\,d\omega \tag{19}$$

where $\vec{F}_{(\alpha)}$ (resp. $\vec{T}_{(\alpha)}$) is the volume force (resp. surface traction vector) given in terms of the covariant basis.

The algorithmic foundation for the implementation of hierarchic shell models based on the displacement formulation is the principle of virtual work (Eq. (10)) with $B(\vec{u}_{(\alpha)},\vec{v}_{(\alpha)})$ and $F(\vec{v}_{(\alpha)})$ given by equations (18) and (19). Dimensional reduction is effected upon integrating with respect to the variable ν.

3.3 Approximability and hierarchic consistency

We state without proof some important theoretical results. Details are available in [7]. We denote the exact solution of the 3D elasticity problem (10) by $\vec{u}_{3D}$ and the exact solution of the hierarchic shell model by $\vec{u}_m$ where the definition of m for thin solid models is given by Eqn. (23) and for the hierarchic shell models it is given by $m = \min(m_\alpha, m_\beta, m_n)$.

1. Approximability: The exact solutions of the hierarchic shell models converge to the exact solution of the problem of three-dimensional elasticity:

$$\lim_{m\to\infty} \frac{\|\vec{u}_{3D} - \vec{u}_m\|_{E(\Omega)}}{\|\vec{u}_{3D}\|_{E(\Omega)}} = 0. \tag{20}$$

2. Asymptotic consistency: As the thickness of the shell is reduced the exact solutions of all hierarchic shell models converge in energy norm to the exact solution of the problem of three-dimensional elasticity:

$$\lim_{t\to 0} \frac{\|\vec{u}_{3D} - \vec{u}_m\|_{E(\Omega)}}{\|\vec{u}_{3D}\|_{E(\Omega)}} = 0. \tag{21}$$

The models (m_α, m_β, m_n) do not satisfy the requirement of asymptotic consistency for $m_n = 0, 1$ but all other models do. We note that convergence occurs in energy norm. Pointwise convergence generally does not occur in the boundary layer [7].

3.4 Classical shell models

Two classical shell models are briefly discussed in the following. These models are not members of the hierarchic sequence of models described in Section 3.2 but rather 'downward' extensions that satisfy the requirements of asymptotic consistency [9].

3.4.1 The Naghdi shell model The Naghdi shell model [10] is as an extension of the Timoshenko beam model and the Reissner-Mindlin plate model to shells: It is assumed that normals to the mid-surface prior to deformation remain straight lines but not necessarily normals after deformation. In other words, the kinematic assumptions account for some transverse shearing deformation. Specifically, the kinematic assumptions are the same as the kinematic assumptions for the hierarchic shell model (1,1,0):

$$\begin{aligned} u_\alpha &= u_{\alpha|0}(\alpha,\beta) + u_{\alpha|1}(\alpha,\beta)\nu \\ u_\beta &= u_{\beta|0}(\alpha,\beta) + u_{\beta|1}(\alpha,\beta)\nu \\ u_n &= u_{n|0}(\alpha,\beta). \end{aligned} \tag{22}$$

The difference between the hierarchic shell model (1,1,0) and the Naghdi shell model is in the material stiffness matrix $[E]$. Whereas in Eqn. (18) the definition of $[E]$ is that of the three-dimensional stress-strain relationship, in the Naghdi model $[E]$ is replaced by its plane stress equivalent. This is necessary for asymptotic consistency, see for example [4].

3.4.2 The Novozhilov-Koiter shell model The Novozhilov-Koiter shell model is an extension of the Euler-Bernoulli beam model and the Kirchhoff plate model to shells: It is assumed that normals to the mid-surface prior to deformation remain normals after deformation [11]. This is known as the Kirchhoff-Love hypothesis. The formulation involves replacement of the field functions u_α and u_β in Eqn. (22) by a linear combination of the first derivatives of u_n. Consequently the number of field functions is reduced to three and the second derivatives appear in Eqn. (18) which implies that the space of admissible functions has to be class C^1. From the point of view of computer implementation the advantages of using only three field functions are far outweighed by disadvantages of the requirement of C^1 continuity and restrictions on the kinematic boundary conditions.

In the ORNL investigation [1], [2] the Novozhilov-Koiter shell model was approximated by an assembly of flat membrane elements and plate elements based on the Kirchhoff plate model.

3.5 Hierarchic 'thin solid' models

An alternative approach to hierarchic semi-discretization is to write the Cartesian components of the displacement vector in the form

$$\begin{aligned} u_x &:= \sum_{i=0}^{m} u_{x|i}(\alpha,\beta)\phi_i(\nu) \\ u_y &:= \sum_{i=0}^{m} u_{y|i}(\alpha,\beta)\phi_i(\nu) \\ u_z &:= \sum_{i=0}^{m} u_{z|i}(\alpha,\beta)\phi_i(\nu) \end{aligned} \tag{23}$$

and the generic form of the virtual work of internal stresses is given by the expression

$$B(\vec{u}_{(x)}, \vec{v}_{(x)}) := \int_\omega \int_{-t/2}^{+t/2} \left([\tilde{D}]\vec{v}_{(x)}\right)^T [E][\tilde{D}]\vec{u}_{(x)}\, d\nu\, d\omega. \tag{24}$$

The generic form of the virtual work of external forces is:

$$\begin{aligned} F(\vec{v}_{(\alpha)}) := &\int_\omega \int_{-t/2}^{+t/2} \vec{F}_{(x)} \cdot \vec{v}_{(x)}\, d\nu\, d\omega + \int_{\partial\Omega} \vec{T}_{(x)} \cdot \vec{v}_{(\alpha)}\, dS \\ &+ \int_\omega \int_{-t/2}^{+t/2} \left([\tilde{D}]\vec{v}_{(x)}\right)^T [E]\{c\}\tau\, d\nu\, d\omega \end{aligned} \tag{25}$$

where $\vec{F}_{(x)}$ (resp. $\vec{T}_{(x)}$) is the Cartesian volume force vector (resp. surface traction vector) given in terms of the curvilinear variables (α, β, ν). This formulation was first discussed in [12].

3.6 Advantages and disadvantages of hierarchic shell and thin solid models

The advantages of thin solid formulations over shell formulations are that they are easier to implement and continuity with other bodies, such as stiffeners, are easier to enforce. The disadvantages are that thin solid formulations cannot be applied to laminated shells unless each lamina is explicitly modeled; the number of field functions must be the same for each displacement component, and the downward extension of the family of hierarchic models to satisfy the requirement of asymptotic consistency cannot be applied. Therefore the anisotropic spaces with $q = 1$ are similar but not equivalent to the Reissner-Mindlin plate model or the Naghdi shell model.

3.7 Adaptivity and feedback

The problem of adaptive selection of dimensionally reduced working models for linear elliptic problems has been addressed in a number of papers, see for example, [13], [14], [15]. The objective was to control the error in energy norm. In practical problems it is necessary to show not only that the error in energy norm is small but also that the data of interest are substantially independent of the choice of a working model, which generally includes testing for the effects of nonlinear terms. For this reason it would be difficult to construct algorithms for automating the selection of working models. The analyst must rely on feedback information developed through a trial and error procedure. Such a procedure is illustrated in this paper.

4 Finite element spaces

Finite element spaces are constructed by partitioning the solution domain Ω into finite elements. A partition will be denoted by Δ, the number of elements of the partition by $M(\Delta)$ and the kth element by Ω_k. Typically Ω_k is mapped from a corresponding standard element Ω_{st} by smooth mapping functions $\mathbf{Q}^{(k)}$. Mapping of the standard hexahedral element $\Omega_{st}^{(h)}$ into Ω_k is illustrated schematically in Fig. 1. The set of all mapping functions is denoted by $\mathbf{Q}$. Typically polynomial basis functions are defined on Ω_{st}. The finite element space $S(\Omega)$ is the span of the mapped basis functions defined on Ω_{st} subject to the continuity requirement of the formulation.

Ideally, the design of finite element spaces is governed by a priori information available concerning the regularity of the exact solution. Practical considerations include the complexity of the solution domain and the capabilities and limitations of the available mesh generators. In the neighborhood of singular points and lines the mesh should be graded in geometric progression toward the singular point and line with a fixed common factor when p-extension is used [4]. If h-extension is used then radical grading is optimal [16]. In the case of plate- and shell-like domains the presence of boundary layers has to be considered.

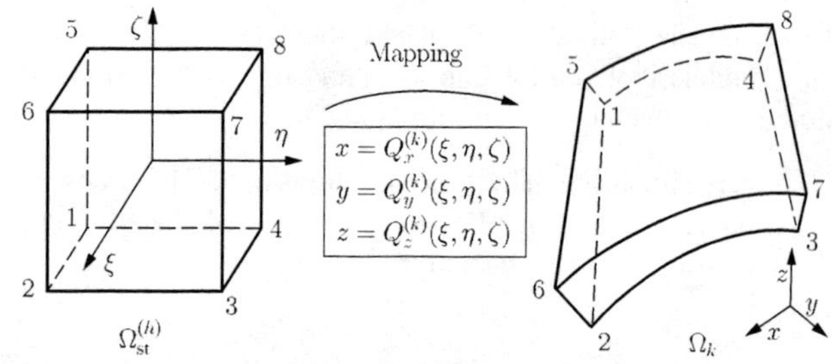

Figure 1. MAPPING OF THE STANDARD HEXAHEDRAL ELEMENT INTO A HEXAHEDRAL ELEMENT.

4.1 The finite element space used in the ORNL investigation

At the time of the ORNL investigation the treatment of shell models by the finite element method was in its very early stages of development. Although the investigators were aware of some contemporary work on curved shell elements, shells were commonly approximated by flat plate elements and most of the available experience was with those elements. For this reason the investigators decided to use flat plate elements for the purpose of analyzing the shell intersection problem [2]. The Hsieh-Clough-Tocher (HCT) triangular element [17] was chosen for the approximation of the displacement component normal to the midsurface of the shell. The constrained linear strain triangle (CLST) was used for approximating the membrane components. A brief description follows.

The HCT triangle is a composite element, comprised of three sub-triangles. On each sub-triangle an incomplete cubic polynomial approximation, comprised of 9 terms, is used. The polynomials are chosen so that along the external edges the normal derivative varies linearly. Therefore there are 9 coefficients per sub-triangle. C^0 continuity is enforced for the sub-elements by constructing basis functions corresponding to the 3 nodal displacements and 6 rotations for each sub-triangle, as indicated in Fig. 2(a) where the circles represent transverse displacements and the arrows represent first derivatives (rotations in the sense of the arrows). The sub-triangles are assembled, which is equivalent to satisfying C^0 continuity over the triangle.

At this point there are 3 internal degrees of freedom, indicated in Fig. 2(a) by the closed circles and arrows, and 9 external degrees of freedom, indicated by the open circles and arrows. In order to satisfy exact and minimal C^1 continuity, the continuity of the normal derivatives along the internal edges is enforced leading to 3 constraint equations that establish a relationship be-

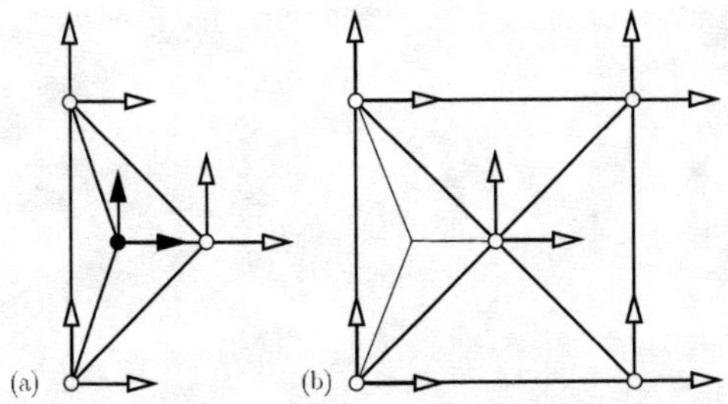

Figure 2. (a) THE 12 DEGREES OF FREEDOM HCT TRIANGLE. (b) COMPOSITE NON-PLANAR QUADRILATERAL ELEMENT ASSEMBLED FROM FOUR HCT TRIANGLES.

tween the internal and external degrees of freedom. Using these constraint equations the internal degrees of freedom are eliminated. Thus the HCT triangle has 9 degrees of freedom: three displacements and three rotations in each coordinate direction.

Non-planar quadrilateral plate elements assembled from four HCT elements were used for approximating the displacement vector components normal to the shell surface. A typical element is shown in Fig. 2(b). Since the center node generally does not lie in the same plane as the vertex nodes, there is a third rotation component, not present in the constituent triangles. A third rotation component is also present in the assembly of the quadrilateral elements into the stiffness matrix, since adjacent elements are generally not co-planar. The usual treatment is that the rotation components are transformed into a Cartesian system, the origin of which lies on the shell surface at the node and one axis is coincident with the normal to the surface. The rotation component in the direction of the normal is usually neglected. This causes various problems. For details we refer to [1], [2], [18]. For a discussions on low order shell elements we refer to [19] and for a benchmark study of cylindrical shell we refer to [20].

The in-plane (membrane) components of the displacement vector were approximated by a similar assembly of triangles. These vector components were approximated over each component triangle by quadratic polynomials constrained so that the variation is linear over the external edges. This is known as the constrained linear strain triangle (CLST). A quadrilateral membrane element has two degrees of freedom at each of its five nodes [2]. The finite element mesh is shown in Fig. 3.

Essentially the shell was represented by an assembly of HCT and CLST triangles. The investigators recognized two sources of error: The errors caused by using plate elements rather than curved shell elements and by using a finite number of elements. Implied is the tacit assumption that as the number of elements is increased the approximate solution will converge to the exact solution of the Novozhilov-Koiter shell model. This has not been proven. Therefore there is no guarantee that the sequence of finite element solutions obtained by h-extension will converge to the exact solution of the Novozhilov-Koiter model.

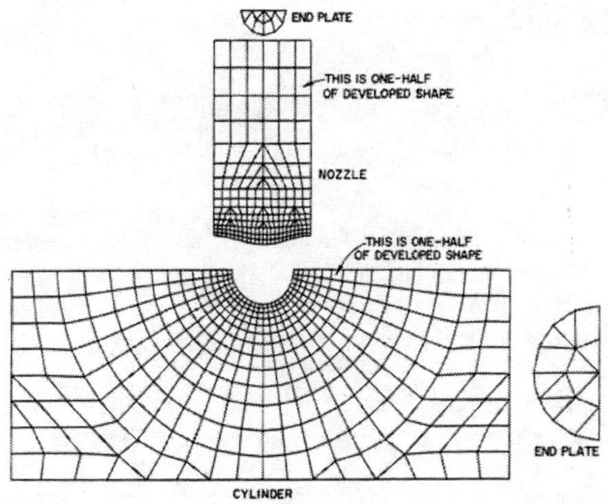

Figure 3. FINITE ELEMENT MESH USED IN THE ORNL INVESTIGATION. THERE ARE 649 NODAL POINTS, OF WHICH 25 NODAL POINTS ARE AT THE INTERSECTION. SOURCE: ORNL-DWG 69-10664R. REPRINTED WITH PERMISSION FROM OAK RIDGE NATIONAL LABORATORY.

Remark 4.1. The definition of the HCT triangle given in [21] is different from the one described here in that all sub-triangles are complete polynomials of degree 3. In this case there are 12 degrees of freedom, the nine degrees of freedom shown in Fig. 2(a) plus the first derivative in the direction of the normal at the midpoint of each side.

4.2 The finite element spaces used in the present investigation

One of the most efficient ways for constructing hierarchic sequences of finite element spaces is by p-extension. A brief description of polynomial spaces on the standard hexahedron is presented here. Additional details are available in [4]. The standard hexahedral element, illustrated in Fig. 1, is defined by

$$\Omega_{\mathrm{st}}^{(h)} := \{(\xi,\eta,\zeta)\,|\,|\xi|<1,\ |\eta|<1,\ |\zeta|<1\}. \tag{26}$$

Each element is mapped from the standard hexahedron defined by Eqn. (26) by smooth mapping functions, as indicted schematically in Fig. 1. For details we refer to [22].

The finite element space $S(\Omega)$ is characterized by the mesh Δ, the mapping functions $\mathbf{Q}$ and the standard polynomial spaces such as those described in the following. The mapping functions establish one-to-one correspondence between the standard element and each element of the mesh. Exact and minimal continuity is enforced so that $S(\Omega) \subset C^0(\Omega)$. For additional details we refer to [4] and [9]. The finite element spaces used in the present investigation are known as anisotropic trunk spaces. For $p > 1$,

$1 \leq q \leq p$ these spaces are defined by:

$$S_{\text{tr}}^{ppq}(\Omega_{\text{st}}^{(h)}) := \text{span}(\xi^k\eta^\ell\zeta^m, \xi^p\eta\zeta^m,\ \xi\eta^p\zeta^m,\ (\xi,\eta,\zeta) \in \Omega_{\text{st}}^{(h)},$$
$$k,\ell = 0,1,2,\ldots,\ k+\ell \leq p,\ m = 0,1,2\ldots,q). \qquad (27)$$

The isotropic trunk space, denoted by $S_{\text{tr}}^{p}(\Omega_{\text{st}}^{(h)})$, is defined in [4].

Various hierarchic and non-hierarchic basis functions have been developed. For details we refer to [4], [23], [24]. For a description of pentahedral elements we refer to [4].

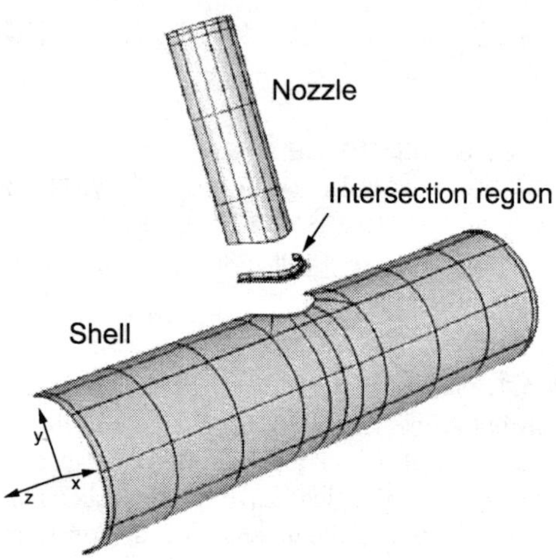

Figure 4. THE 188-ELEMENT MESH USED IN THE PRESENT INVESTIGATION.

5 Experiments

The test article was made by welding two carbon steel pipes then carefully machining the weldment to the test dimensions. The test article was annealed at several points in the machining process. The experimental arrangement is shown in Fig. 5. The horizontal part is called the cylinder, the vertical part is called the nozzle. The length of the cylinder was 39.0 in (991 mm). The length of the nozzle, measured from the point of intersection of the centerline of the nozzle with the centerline of the cylinder was 19.5 in (495 mm). The outside diameter of the cylinder (resp. nozzle) was 10 in (254 mm) (resp. 5.0 in (127 mm)). The intended wall thickness of the cylinder (resp. nozzle) was 0.1 in (2.54 mm) (resp. 0.05 in (1.27 mm)).

5.1 Constraint conditions

As shown in Fig. 5, the right end of the cylinder was rigidly clamped to a heavy flat plate bolted to a frame. Small flanges were machined into the ends of the cylinder and nozzle to support the seal and the clamping forces. Heavy loading fixtures were attached on the left end of the cylinder and the end of the nozzle to provide seal and seating for the application of forces.

Figure 5. EXPERIMENTAL ARRANGEMENT. REPRODUCED WITH PERMISSION FROM OAK RIDGE NATIONAL LABORATORY.

5.2 Loading conditions

A total of thirteen (13) load cases were investigated. These included pressure loading and axial forces, shear forces and moments applied to the cylinder and the nozzle through hydraulic rams acting through load cells. For all 13 load cases the load was applied in increments of 20 percent of the full load, then decreased to zero again in 20 percent decrements. In this paper only one of the loading cases, the pressure load, is discussed. Additional load cases are discussed in [25]. The maximum value of the pressure was 50.0 psi (344.8 kPa).

The pressure loading was applied by means of a hydraulic fluid. In order to compensate for the weight of the hydraulic fluid, a counterbalancing force was applied to the fixture at the free end of the cylinder through a cable that is visible in Fig. 5.

5.3 Measurements

Details of the instrumented intersection region is shown in Fig. 6.

A total of 322 three-gauge (Micro-Measurements type EA-06-030YB-120, option SE) foil rosettes were bonded on the inside and outside surfaces by epoxy adhesive and cured. The gauges in the rosettes were arranged in a 'Y' pattern (i.e., the directions of measurement were 120 degrees apart) [2]. These gauges consist of a grid of very fine wire bonded to a foil backing. The electrical resistance of the grid varies linearly with strain. The strain is found by measuring the change of resistance in the wires.

The ratio of the resistance change in a strain gauge to the Lagrangian strain causing the change is called the gauge factor. The gauge factor of each production lot is determined by sample measurements and is given on each package with its tolerance. Typical tolerances are 0.5 to 1.0 %. No tolerance data are provided in references [1], [2].

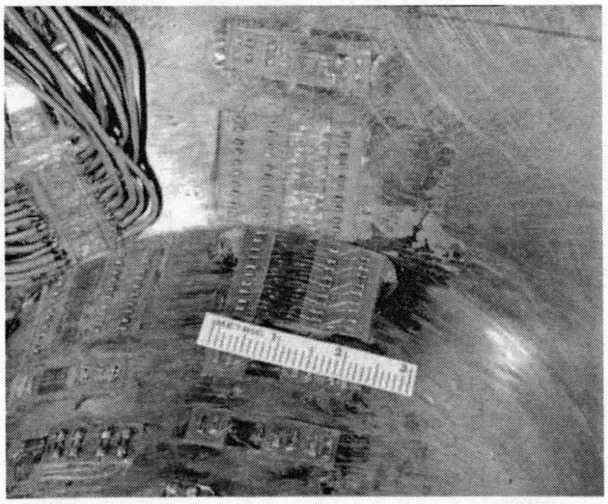

Figure 6. DETAIL OF TEST ARTICLE INSTRUMENTED WITH STRAIN GAUGES. REPRODUCED WITH PERMISSION FROM OAK RIDGE NATIONAL LABORATORY.

6 Working models

The quality of a working model is determined by the proximity of its exact solution to that of the state of the art model, more precisely, the proximity of functionals that would be computed from the exact solution of the state of the art model and the working model described in Section 2. Here we examine a sequence of working models from the point of view of their ability to approximate physical observations and data computed from physical observations through simple transformation.

The objective of the ORNL experiments was to investigate how well thin shell theory is capable of predicting the stress distribution in intersecting cylindrical shells in the neighborhood of the intersection. The investigators had in mind the Novozhilov-Koiter shell model only. Very likely the Novozhilov-Koiter model is a reasonable choice for representing the smooth parts of the intersecting shells, but the assumptions incorporated in this model are not valid at the intersection. Whereas the mathematical problem of the intersecting cylinders is well defined [26], it is not a good representation of the physical problem.

The working models employed in the present investigation were also based on the theory of elasticity but differ from the Novozhilov-Koiter shell model with respect to kinematic assumptions and material properties, as described in Section 3. The intersection region was treated as a three-dimensional elastic region for each model. The size of the intersection region is fixed for each working model. It is illustrated in Fig. 7. Elsewhere the kinematic assumptions of the thin solid formulation were used with $q = 1, 2, \ldots$

The problem of selecting a working model with respect to the particular goals of computation of the ORNL experiments is understood as follows: A particular working model based on the thin solid formulation is not rejected if the equivalent (von Mises) stresses computed in the gauge locations do not differ by more than 2 percent from the corresponding equivalent stresses computed from the fully three-dimensional model. The choice of 2 percent tolerance is, of course, arbitrary, however the accuracy required of the numerical solution depends on this choice. The relative error in the numerical solution has to be less than approximately 1 percent.

Using the 188-element mesh shown in Fig. 4 and the anisotropic spaces $S_{\mathrm{tr}}^{ppq}(\Omega_{\mathrm{st}}^{(h)})$ with $q = 1, 2, \ldots, 8$ all working models yielded consistent results within the prescribed tolerances. Therefore the simplest model ($q = 1$) is preferable.

In comparing data computed from working models with physical measurements, it is necessary to recognize the differences between the mathematical problem being solved and the physical system being modeled. These differences are enumerated in the following in relation to the ORNL test article.

1. Geometric variations. The ORNL investigators were careful to minimize errors in manufacturing the test article however, owing to unavoidable machining tolerances, some variations in wall thickness and the other dimensions had to be present. Quoting from reference [2];

   "A careful dimensional inspection of the machined model indicated that, despite the care taken in machining, there were wall thickness variations in both nozzle and cylinder with the nozzle thickness being as much as 15 % greater (0.007 to 0.008 in. compared with the nominal 0.050 in) in the fourth quadrant than in the second."

   In the working models it is assumed that the shells are defined by perfect cylindrical surfaces and constant wall thickness.

   The intent of the ORNL investigators was to manufacture the intersection with zero fillet radius. In reality, the milling tool leaves some fillet. In the working models the fillet radius is zero.

   The test article had small flanges at the ends. The working models do not account for those flanges.
2. Variations in material properties. The material properties

assumed by the ORNL investigators are the nominal elastic constants of carbon steel. Modulus of elasticity: $E = 30 \times 10^6$ psi (207 GPa); Poisson's ratio: $\nu = 0.3$. The actual elastic constants of the test article can differ from the nominal values by a few percent. There are no data on the statistical variations of the modulus of elasticity and Poisson's ratio, however it is reasonable to expect that the mean value of E (resp. ν) is within about 2 % (resp. 5 %) of the nominal value. Therefore systematic as well as random errors are present in making comparisons between measured strains and strains computed from the working models.
The relationship between stress and strain in the test article will become nonlinear when the strain corresponding to the proportional limit is exceeded. In the working models examined herein the material is assumed to be perfectly elastic independent of the magnitude of strain.

3. Differences in constraint conditions. Rigid end fixtures were attached to the free end of the cylinder and the nozzle, as described in Section 5.1. Details on the end fixtures are not given in the ORNL reports, however the investigators assumed that the end fixtures were sufficiently rigid to constrain the ends of the cylinder and nozzle so as to maintain the ends as plane circles [2]. In the working model constructed by the ORNL investigators the end fixtures were represented by end plates. In the present investigation the fixture attached to the nozzle was represented by an end plate. At the free end of the cylinder the radial and tangential displacement components were set to zero.
4. Differences in loading conditions. The test article was loaded through hydraulic rams acting on the end fixtures. The accuracy of the applied load and hence the accuracy of the stress resultants was determined by the accuracy of the load cells. The transfer of the load through the end fixtures was through mechanical contact. The precise distribution of the tractions acting on the ends is not known. In the working models used in the present investigation uniform tractions were applied on the free end of the cylinder.

It is seen that even under very carefully controlled experimental conditions some degree of uncertainty concerning the physical system is present. Some of these uncertainties can be reduced, others either cannot be reduced or may not be feasible to reduce. For example, the mean value of the elastic constants can be determined by simple coupon tests. The dimensions of the test article can be measured with high accuracy. On the other hand, it would be very difficult to determine the distribution of the tractions or constraint conditions imposed by the end fixtures. In addition, some degree of uncertainty is associated with the instruments employed in making the observations and the effects of the environment on the instruments.

In view of these uncertainties one cannot expect very close correlation between computed and experimental data. The largest uncertainties are caused by the difficulties associated with manufacturing thin-walled objects to tight tolerances and mathematical representation of constraint conditions.

7 Numerical results

The results presented herein are representative of the data that can be readily obtained by finite element analysis. It is necessary to ascertain that the errors in the computed data are well below the threshold set for rejection of a working model. In this investigation the following steps were taken: (a) The relative error in energy norm was estimated. This provides an overall view of the quality of the approximation. The error in energy norm is roughly equivalent to the RMS error in stresses [4]. For the 188-element mesh the estimated relative error in energy norm ranged between 1.15 and 3.20 percent. (b) The dependence of the data of interest on the mesh and the polynomial degree of elements was examined. It was found that the data of interest are substantially independent of the mesh and the polynomial degree of elements. (c) The data of interest were examined for jump discontinuities at inter-element boundaries. Substantial jump discontinuities in locations where the data should be continuous is an indication that the discretization is inadequate. Significant discontinuities were not found.

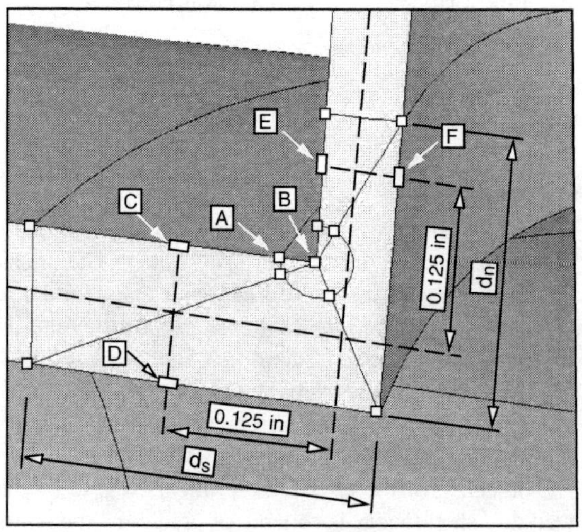

Figure 7. MESH DETAIL AT THE INTERSECTION REGION AND LOCATION OF THE STRAIN GAUGES IN THE PLANE OF SYMMETRY NEAREST TO THE INTERSECTION. 188-ELEMENT MESH.

Due to space limitations, only some of these procedures are illustrated in the following. Details of the finite element mesh in the shell intersection region are shown in Fig. 7. The size of the shell intersection region is characterized by the dimensions d_s

and d_n. Unless otherwise stated, in the numerical investigation described herein these dimensions were fixed: $d_s = 0.25$ in (6.35 mm); $d_n = 0.20$ in (5.08 mm).

The layout of the mesh in the intersection region is typical of meshes used in p-extensions when corner singularities are present [4]. Nodal point B lies on the line of intersection between the outer surface of the nozzle and the outer surface of the shell in the plane of symmetry. Nodal point A lies on the line of intersection between the outer surface of the shell and the plane of symmetry.

The convergence of the equivalent stress at Point A with respect to the number of degrees of freedom N is shown in Fig. 8 on a semi-log scale for the fully three-dimensional model. It is seen that the stress is substantially independent of the polynomial degree p for $p \geq 5$. This is a necessary criterion for ensuring that the equivalent stress computed from the finite element solution is close to its exact value.

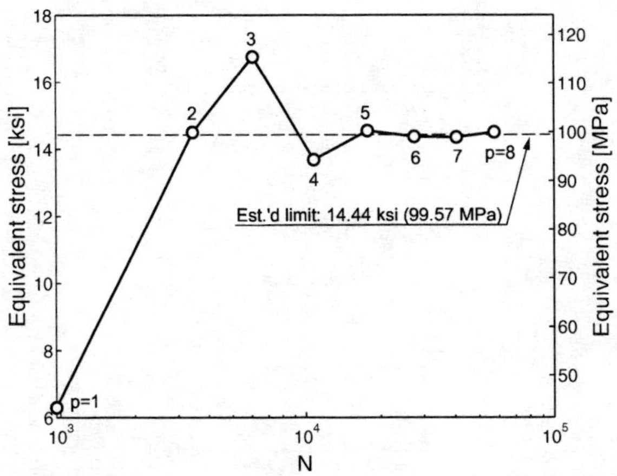

Figure 8. CONVERGENCE OF THE EQUIVALENT STRESS AT NODE A SHOWN IN FIG. 7. 188-ELEMENT MESH.

A similar plot shown in Fig. 9 for the equivalent stress computed for point B clearly indicates divergence. This is caused by the presence of an edge singularity. The equivalent stress computed for the exact solution of the elasticity problem is infinity in this point, hence the equivalent stress computed from the finite element solution cannot converge to a finite value. Consequently extrapolation of strain data from the gauge locations to the intersection is not permissible.

Equivalent stresses in strain gauge locations nearest to the intersection region in the plane of symmetry (toward the fixed-end of the cylinder) are shown in Table 1. These data were computed from the solutions obtained for the fully three-dimensional model with p ranging from 1 to 8. The number of degrees of freedom (N) are shown in the second column. Points C, D, E

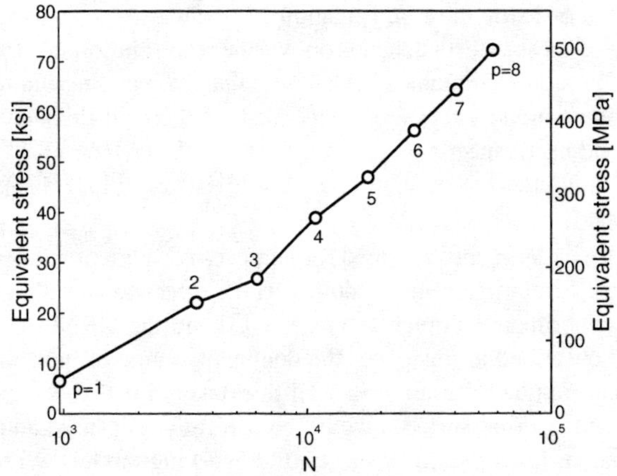

Figure 9. DIVERGENCE OF THE EQUIVALENT STRESS AT NODE B SHOWN IN FIG. 7. 188-ELEMENT MESH.

and F are shown in Fig. 7. Each stress value converges strongly to a limit value, similar to the convergence shown in Fig. 8. The corresponding experimental data and the relative differences (DIFF), using the computed values as the base, are shown in the last two rows. It is seen that the differences are very substantial in three of the four points.

Table 1. EQUIVALENT STRESSES (PSI) IN THE GAUGE LOCATIONS C, D, E, F IDENTIFIED IN Fig. 7. FULLY THREE-DIMENSIONAL MODEL. THE POINTS ARE LOCATED ON THE FIXED-END SIDE OF THE CYLINDER (1 PSI=6.895 kPa).

p	N	Pt. C	Pt. D	Pt. E	Pt. F
3	6095	11225	11959	13729	16589
4	10828	12461	11680	16729	19687
5	17774	12455	12158	17644	19441
6	27497	12535	12089	17237	19005
7	40561	12542	12021	17108	19104
8	57530	12544	11982	17094	18903
EXPT		15569	14554	16981	13100
DIFF. (%)		28.6	24.1	-0.3	-28.9

These differences are attributed to variations in wall thickness. All other uncertainties would have a much smaller effect. Inspection of the test article revealed variations in wall thickness as large as 15 % [2]. Variations in wall thickness affect both the distribution and magnitude of stresses. The magnitude of these effects depends on whether the solution is bending or membrane dominated. Assuming that the bending moments and membrane forces are independent of the wall thickness, in a bending (resp. membrane) dominated region 10% change in wall thickness causes approximately 24% (resp. 11%) change in stress.

In experiments designed for purposes of validation it is necessary to eliminate uncertainties with respect to the object of the experiment as much as possible [3]. In the ORNL test article described in this paper the dominant source of uncertainty is the unknown variation in wall thickness. Ideally, the external and internal surfaces would be carefully measured and the actual surfaces would be used in the working model. With today's technology it would be possible to produce a CAD model of the test article as manufactured, to within 0.001 in (0.025 mm) tolerance. The complexity of the resulting geometrical description would make construction of the working model significantly more complicated, however.

In the immediate vicinity of the junction there are large bending moments that decay with respect to distance from the junction. The computed and experimentally obtained values of the equivalent stress in gauge locations along the line of intersection of the plane of symmetry with the outer surface of the cylinder on the fixed end side are tabulated in Table 2. It is seen that the errors are larger near the junction than away from it, and the errors are roughly consistent with approximately 10% variation in wall thickness. The magnitude of the actual variation is unknown.

The other sources of error are inelastic deformation in the vicinity of the junction, variations in material properties, errors in the location of the strain rosettes, errors in measurement, errors caused by idealization of the actual constraint condition and errors in loading. None of these errors are large enough to explain the observed differences. For example, to test the effect of inelastic deformation, elastic perfectly plastic material behavior was assumed, the yield stress being 36.0 ksi (248 MPa). Using the deformation theory of plasticity it was found that the plastic zone was confined to a very small neighborhood of the line of intersection of the outer surfaces and the effect on the stresses in the gauge locations was negligible. For a discussion on the other effects we refer to [25].

A comparison between data computed from experimental measurements and data computed from the fully three-dimensional working model along the inner surface of the cylinder, measured from the mid-surface of the nozzle, is shown in Fig. 10.

Equivalent stresses computed in the strain gauge locations labelled C, D, E, F in Fig. 7 (see also Table 1) from solutions obtained by means of the hierarchic thin solid models based on the spaces $S_{\mathrm{tr}}^{ppq}(\Omega_{\mathrm{st}}^{(h)})$ are shown in Table 3. The estimated relative errors in energy norm, denoted by $(e_r)_E$, are also given in Table 3. These estimates are based on p-extension. Details are available in [4]. It is seen that the stress data are insensitive to the choice of thin solid model.

One of the modeling assumptions is the selection of the size of the intersection region indicated in Fig. 7 by labels d_s and d_n. The data presented so far was computed with $d_s = 0.25$ in (6.35

Table 2. COMPUTED (FEA) AND EXPERIMENTALLY (EXP) OBTAINED VALUES OF THE EQUIVALENT STRESS (PSI) IN GAUGE LOCATIONS ON THE INTERSECTION OF THE PLANE OF SYMMETRY WITH THE OUTSIDE SURFACE OF THE CYLINDER AS A FUNCTION OF THE DISTANCE (s) FROM THE MID-SURFACE OF THE NOZZLE (INCHES).

s	FEA	EXP	err (%)
0.125	12541	15569	24.14
0.250	10129	12381	22.23
0.375	8190	8870	8.31
0.500	6531	7453	14.13
0.625	5086	5783	13.70
1.000	2105	2242	6.49
1.500	1836	1844	0.42
2.000	2343	2230	-4.84
3.000	2431	2292	-5.71

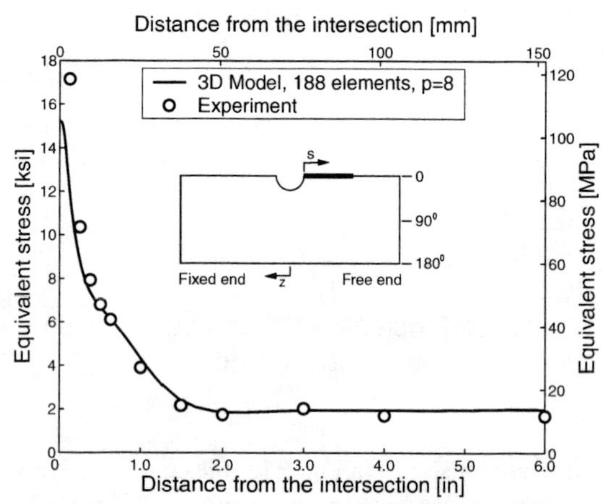

Figure 10. EQUIVALENT STRESS (σ_{eq}) ON THE INSIDE SURFACE OF THE CYLINDER VS. DISTANCE FROM THE MID-SURFACE OF THE NOZZLE.

Table 3. EQUIVALENT STRESSES (PSI) IN THE GAUGE LOCATIONS C, D, E, F IDENTIFIED IN Fig. 7. THIN SOLID MODELS. THE POINTS ARE LOCATED ON THE FIXED-END SIDE OF THE CYLINDER (1 PSI=6.895 kPa).

$d_s = 0.25$ in (6.35 mm), $d_n = 0.2$ in (5.08 mm)							
p	q	$(e_r)_E$	N	Pt. C	Pt. D	Pt. E	Pt. F
8	1	3.20	41032	12110	11730	16926	18432
8	2	1.15	44766	12570	11978	17094	18904
8	3	1.35	47832	12552	11979	17094	18904
8	8	1.35	57530	12544	11982	17094	18903

Table 4. EQUIVALENT STRESSES (PSI) IN THE GAUGE LOCATIONS C, D, E, F IDENTIFIED IN Fig. 7. THIN SOLID MODELS. THE POINTS ARE LOCATED ON THE FIXED-END SIDE OF THE CYLINDER (1 PSI=6.895 kPa).

$d_s = 0.13$ in (3.30 mm), $d_n = 0.15$ in (3.81 mm)							
p	q	$(e_r)_E$	N	Pt. C	Pt. D	Pt. E	Pt. F
8	1	2.44	41032	12111	12470	17076	18169
8	2	1.11	44766	12621	12026	17104	19141
8	3	1.32	47832	12564	11941	17104	19143
8	8	1.31	57530	12546	11964	17106	19143

mm) $d_n = 0.20$ in (5.08 mm). Therefore points C, D, E, F were located within the intersection region. Letting $d_s = 0.13$ in (3.30 mm) $d_n = 0.15$ in (3.81 mm) the points are located in the thin solid region. Repeating the computations the results shown in Table 4 are obtained. On comparing Table 3 with Table 4 it is seen that the differences are small, indicating that the working models are insensitive to the size of the intersection region.

8 Summary and conclusions

The intent of the ORNL investigation was to determine whether the finite element space characterized by the mesh shown in Fig. 3 and the HCT plate elements, combined with the CLST plane stress elements, are capable of approximating strains in the given gauge locations. We understand the problem to consist of three parts: (a) Whether the Novozhilov-Koiter model is capable of representing the deformation of the test article in the intersection region; (b) whether the three-dimensional assembly of planar plate and membrane elements converge to the exact solution of the Novozhilov-Koiter model as the size of the elements is reduced, and (c) whether the mesh shown in Fig. 3 is suitable for controlling the errors of discretization in the sense of equation (4). The ORNL investigation did not address these questions separately.

In the present investigation the need to address these questions was avoided by (a) demonstrating that hierarchic thin solid formulations provide consistent results in the gauge locations even for the lowest member of the hierarchy, characterized by $q = 1$; (b) curved elements mapped by smooth functions rather than planar elements were used, and (c) the data of interest were shown to be substantially independent of the discretization. Another important difference between the present investigation and the ORNL investigation is that in the present investigation the intersection region was exempt from the kinematic assumptions of the thin solid formulation.

The ORNL investigation highlights some of the difficulties and limitations of experimental validation of working models for thin shell intersection problems. The experimental data are dominated by uncertainties caused by difficulties associated with the fabrication of thin-walled objects to exacting tolerances. Increasing the wall thickness would reduce errors caused by manufacturing tolerances but then the thin shell model would not be applicable.

The primary objective in choosing a working model is to account for all physical laws and relationships that have a significant influence on the data of interest. Therefore the choice of a working model depends on the data of interest and it has to be validated with respect to the data of interest. The secondary objective is to identify the simplest working model suitable for the computation of the data of interest.

In correlation with experimental observations a working model is tested against data that are either observable or can be computed from observable data. In the model problem discussed herein the strain components in surface points located in the vicinity of the shell intersection were measured and the von Mises stresses were computed in the gauge locations.

The data of interest are generally not observable, however. For example, one may be interested in the maximum value of the integral of the normal stress over some small area, a functional that cannot be observed. Therefore even if a working model were shown to be successful in predicting certain measured data, it may not be suitable for computing other data of interest. It is necessary to have means for systematic evaluation of the effects of various assumptions incorporated in a working model on the data of interest. Hierarchic modeling is an essential tool for validating a working model with respect to the data of interest.

The validity of a working model cannot be established by experimental correlation in general. The purpose of validation experiments is to determine whether certain necessary conditions are met by a working model. Validation experiments cannot establish sufficient conditions for acceptance.

ACKNOWLEDGMENT

The authors acknowledge with thanks partial support received from the Air Force Office of Scientific Research under Grant No. F49620-01-1-0074. The authors thank Oak Ridge National Laboratory for the copies of reports and photographs that document the physical experiments and numerical investigation of the shell intersection problem discussed herein. Dr. Ricardo Actis of Engineering Software Research and Development, Inc. provided valuable advice on the uses of StressCheck in connection with this investigation.

REFERENCES

[1] R. C. Gwaltney, J. M. Corum, S. E. Bolt, and J. W. Bryson. Experimental stress analysis of cylinder-to-cylinder shell models and comparisons with theoretical predictions. *Journal of Pressure Vessel Technology*, pages 283–289, November, 1976.

[2] J. M. Corum, S. E. Bolt, W. L. Greenstreet, and R. C. Gwaltney. Theoretical and experimental stress analysis of ORNL thin-shell cylinder-to-cylinder Model No. 1. Technical Report ORNL 4553, Oak Ridge National Laboratory, Oak Ridge, Tennessee, October, 1972.

[3] W. Oberkampf. What are validation experiments? *Experimental Techniques*, pages 35–40, May/June 2001.

[4] B. Szabó and I. Babuška. *Finite Element Analysis*. John Wiley & Sons, Inc., New York, 1991.

[5] M. Vogelius and I. Babuška. On a dimensional reduction method I. The optimal selection of basis functions. *Mathematics of Computation*, 37:31–46, 1981.

[6] C. Schwab. *Dimensional Reduction for Elliptic Boundary Value Problems*. PhD thesis, University of Maryland, College Park, MD 20742, 1989.

[7] I. Babuška and L. Li. The problem of plate modeling: Theoretical and computational results. *Comp. Methods Appl. Mech. Engrg.*, 110:249–273, 1992.

[8] R. Actis, B. Szabó, and C. Schwab. Hierarchic models for laminated plates and shells. *Comp. Methods Appl. Mech. Engrg.*, 172:79–107, 1999.

[9] C. Schwab. *p- and hp-Finite Element Methods*. Oxford University Press, Oxford, 1998.

[10] P. M. Naghdi. The theory of shells and plates. In S. Flügge, editor, *Encyclopedia of Physics*, pages 425–640. Springer-Verlag, Berlin, 1972.

[11] V. V. Novozhilov. *Thin Shell Theory*. P. Noordhoff Ltd., Groningen, 2nd edition, 1964.

[12] B. Szabó and G. Sahrmann. Hierarchic plate and shell models based on p-extension. *Int. J. Num. Meth. Engrg.*, 26:1855–1881, 1988.

[13] M. Vogelius and I. Babuška. On a dimensional reduction method III. A posteriori error estimation and an adaptive appproach. *Mathematics of Computation*, 37:361–384, 1981.

[14] I. Babuška,, I. Lee, and C. Schwab. On the a posteriori estimation of the modelling error for the heat conduction in plates and its use for adaptive hierarchic modelling. *Appl. Numer. Math.*, 14:5–21, 1994.

[15] J. T. Oden and J. R. Cho. Adaptive hpq-finite element methods of hierarchical models for plate- and shell-like structures. *Comp. Methods Appl. Mech. Engrg.*, 136:317–345, 1996.

[16] I. Babuška and T. Strouboulis. *The Finite Element Method and its Reliability*. Oxford University Press, Oxford, 2001.

[17] R. W. Clough and J. L Tocher. Finite element stiffness matrices for analysis of plate bending. In *Proceedings of Conference on Matrix Methods in Structural Mechanics, Report AFFDL-TR-66-80*, pages 515–545. Wright-Patterson Air Force Base, Ohio, 1966.

[18] O. Greste. Finite element analysis of tubular K joints. Technical Report UCSESM 70-11, University of California at Berkeley, June, 1970.

[19] J. Pitkäranta. Mathematical and historical reflections on the lowest order finite element models for thin structures. *Computers and Structures*, 81:895–121, 2003.

[20] J. Pitkäranta, Y. Leino, O. Ovaskainen, and J. Piila. Shell deformation states and the FEM. Benchmark study of cylindrical shells. *Comp. Methods Appl. Mech. Engrg.*, 128:81–121, 1995.

[21] P. G. Ciarlet. *The Finite Element Method for Elliptic Problems*. North-Holland, Amsterdam, 1978.

[22] G. Királyfalvi and B. A. Szabó. Quasi-regional mapping for the p-version of the finite element method. *Finite Elem. Anal. Des.*, 27:85–97, 1997.

[23] A. Düster, H. Bröker, and E. Rank. The p-version of the finite element method for three-dimensional curved thin walled structures. *Int. J. Num. Meth. Engrg.*, 52:673–703, 2001.

[24] M. Melenk, K. Gerdes, and C. Schwab. Fully discrete hp-finite elements: Fast quadrature. *Comp. Methods Appl. Mech. Engrg.*, 190:4339–4369, 2001.

[25] D. Muntges. *Validation of working models for thin cylindrical shells*. MS thesis. Washington University, The Henry Edwin Sever Graduate School, St. Louis, MO 63130, 2004.

[26] G. Geymonat and E. Sanchez-Palencia. On the rigidity of certain surfaces with folds and applications to shell theory. *Arch. Rational Mech. Anal.*, 129:11–45, 1995.

Proceedings of IMECE04
2004 ASME International Mechanical Engineering Congress and Exposition
November 13-20, 2004, Anaheim, California USA

IMECE2004-60380

STRESS INTENSITY FACTOR DISTRIBUTION AND CRACK GROWTH IN PRESSURE VESSEL PROBLEMS BY THE FROZEN STRESS METHOD

C. W. Smith[1] and C. T. Liu[2]

[1] Department of Engineering Science and Mechanics
Virginia Polytechnic Institute and State University
Blacksburg, Virginia 24061

[2] Air Force Research Laboratory, PRSM
10 E. Saturn Blvd.
Edwards AFB, California 93524-7680

ABSTRACT

This paper describes the application of a laboratory based experimental method [1] to three dimensional cracked body problems in pressure vessels in order to determine the crack shape and stress intensity factor (SIF) distribution along the crack front when the crack shape is not known a-priori. Results for specific problems are presented and conditions and limitation of the method are described.

INTRODUCTION

Back in the mid sixties, when an explosion of three dimensional fracture problems solutions developed along with the digital computer, the writer and his associates decided that a laboratory based experimental method should be developed in order to partially fill the gap produced by the lack of proof testing for many large structures, such as ships, bridges, aircraft, missiles and pressure vessels because of the excessive costs involved. By effecting a marriage between the equations of linear elastic fracture mechanics (LEFM) with frozen stress photoelasticity (Appendix A), such a method has been developed. Over the years, it has been used to check many numerical solutions in a cost effective manner and, in the process has provided some novel, or unique aspects. The method of analysis with equations for all three local modes of stress intensity is described in [1], and is also presented in [2]. The unique feature lies in the fact that fact that starter cracks may be grown above critical temperature by applying service loads before cooling the model. When the crack reaches its desired size and shape (controlled only by loads and body shape) the loads are reduced in order to stop crack growth and stress freezing is then completed. If sufficient growth is permitted to occur, only Mode I will exist along the crack front. That is, as described by Cotterell [3], [4] the crack is a Class I crack, and by using data sufficiently close to the crack tip, a two term expression for the local stresses becomes adequate for analysis. The resulting equations for both mixed and pure Mode I are presented in Appendix

Approved for public release; distribution unlimited.

B.

EXPERIMENTAL RESULTS

Nozzle Corner Cracks in Reactor Vessels

The writer and his associates were approached by researchers at Delft Technical University in 1976 to evaluate a test developed at Delft for determining stress intensity factors for a nozzle corner crack. They used a flat plate with a nozzle located in its center with a starter crack oriented normal to an external field of uniaxial tension [5]. They grew the crack in a steel model using high cycle tension-tension fatigue loads. The result is shown in Fig 1a for small and moderately deep cracks and in Fig 1c for a deep crack. Using monotonic loads, we produced shapes shown in Figs 1b and 1d which overlaid those obtained at Delft.

We subsequently applied our method to study a similar local geometric situation where the nozzle was located in a scale model of a nuclear pressure vessel. (Fig. 2). The model consisted of two semicylindrical center sections, each containing a nozzle and two hemispherical end domes. Starter cracks were oriented as shown in Fig. 3a and the model was loaded by internal pressure. The crack shapes and normalized SIF distributions obtained are shown in Fig. 3b and Fig. 3c respectively. Also shown in Fig. 3b are the loci of fitted ellipses used by analysts in this problem. Fig. 3c shows how the flattening of the crack shapes near their mid points effectively inverts the SIF distribution along the crack front. When a starter crack was inserted into several models at the $\theta = 45°$ location, (Fig. 3a) very different results occurred. First, the planar starter crack (shaded in Fig. 4a) grew in its plane near n to n'. However, it followed a gently curved path from m to v. The normalized SIF distributions initially showed a domed distribution with peak values near $\beta/\beta_{max} \approx 0.5$ (Fig. 4b). However, between $a'/T' = 0.291$ and 0.458, a crease in the crack surface between the planar and curved parts of the crack surface smoothed out and the SIF distribution inverted with the maximum SIF values occurring at $\beta/\beta_{max} \approx 1.0$ (Fig. 4b). It is noted here that the near tip fringe patterns along the curved portion of the crack were not symmetrical with respect to the crack surface, but no rotation of the fringe loops was present to indicate a shear mode. The lack of symmetry in the fringe loops was due to a varying non-singular tensile stress parallel to the crack growth direction and was not indicative of a shear mode [6]. Generally, the shape taken by the growing crack is the one which minimizes the gradient in the SIF along the flaw border.

Cracks in Rocket Grain

In all of the cases discussed above the starter cracks were, in fact, Class I Cracks. Since, however, they are introduced by striking a sharp blade held normal to a surface, and the resulting crack emanates from the blade tip, the loading for the starter crack is a dynamic wedging force and the induced stress field may differ from internal pressure. This might occur if one drives the starter crack too far in. If the load is then changed to internal pressure, the new load may produce a shear mode at the starter crack tip. In such cases, the crack will then turn in order to eliminate the shear mode as it grows. An example of this type of behavior follows. Figure 5 shows a 1/10 scale model of the crossection of a cylindrical motor grain model. Two off axis (Fig. 5) cracks were introduced in different models but in identical loci of each of the models. Plan and center section SS views of the cracks are shown in Fig. 6. The crack in Model 4 has not grown as far as the one in Model 8. If one assumes that the crack in Model 8 represents a later stage in the growth of the crack in Model 4, one can construct the following scenario. Table I gives pertinent data from the frozen stress tests on the two cracks. The normalized (SIFs) F_1 were computed using an approximation of an elliptical integral of the second kind $\sqrt{Q}$, which amounts to assuming that the cracks were planar semi-elliptic shapes. While this assumption does in fact hold for symmetric cracks (Fig. 5) which enjoy both geometric and load symmetry, such is not

Approved for public release; distribution unlimited.

the case of the "off axis" cracks pictured in Fig. 6, as noted by the direction change in the crack fronts shown in the section SS cuts. The influence of these direction changes, while accommodated in the experimental data, was ignored in computing the values of $\sqrt{Q}$ for use in determining F_i. The form of $\sqrt{Q}$ was developed by Newman [7]. We note that, in our scenario, Model 4 represents an earlier state in the growth of the off-axis crack and that Model 8 indicates a larger crack. Model 4 exhibited both F_1 and F_2 values all along the crack front except at the fin boundary where only F_1 existed, and section SS shows the crack front turning. In Model 8, however, we see a series of river marks on both sides of the crack suggesting the presence of Mode III. No data were taken in these regions however, since it was conjectured that the limitations of linear elastic fracture mechanics may have been exceeded. At all other locations along the Model 8 crack front, however, only pure Mode I was measured suggesting that the crack had grown sufficiently to eliminate all shear modes. The section SS in Model 8 suggests that the turn in Model 4 (which was accompanied by Mode II) has been completed. This means that the crack in Model 8 has achieved, or nearly achieved, Class I status with turning completed and will now proceed in the new direction indicated by Section SS (Fig. 6). It should also be noted that in Model 8, a slight dimple appears in the crack front in the areas of the river markings, suggesting that the effect of shear modes is to retard temporarily the growth of the crack.

Such a scenario as described above leads to an interesting conclusion. Both frozen stress experiments and finite element analysis point to the fin corner as the highest stressed point in the uncracked loaded model. However, any crack initiating there will contain shear modes. On the other hand, a crack initiating on the fin axis (i.e. the symmetric crack in Fig. 5) will show no evidence of shear modes (Fig. 7) and is a Class 1 crack when it appears. As such, one expects such a crack to be far more mobile and dangerous than one emanating from the fin corner. Thus, one cannot predict a fracture on the basis of uncracked model stress analysis alone. According to Cotterell, all Class II cracks (ie here those under shear modes) eventually become Class I cracks and this concept is borne out by these results. The symmetric crack shown in Fig. 5 is already a Class I crack due to absence of shear modes and will not turn during growth.

On the basis of the examples presented above (and other cases studied), we conclude that stable crack growth in pressure vessels proceeds under pure Mode I. Moreover, cracks grown above critical temperature during the stress freezing process in photoelastic models appear to overlay those produced in steel structures by high cycle tension-tension fatigue when test scales are identical.

Table 1

Loads[1]	Crack Description[2] (dimensions in mm)	[3] $F_i = K_i\sqrt{Q}/P_{sf}\sqrt{\pi\alpha}$ i = 1, 2			
		Depth (d)	Surface (s)		
			Top	Bottom	Fin tip surface
$P = 88.97$ N $P_{max} = 0.049$ MPa $P_{sf} = 0.035$ MPa	Model 4 Off-axis inclined $a = 8.71$ $\Delta a = 2.18$ $c = 11.15$ $\Delta c = 3.02$ $a/c = 0.78$ $a/t = 0.23$	$F_1 = 2.09$ $F_2 = 0.53$	[4]ns 20° 2.88*	ns 20° 2.81*	
$P = 88.97$ N $P_{max} = 0.103$ MPa $P_{sf} = 0.049$ MPa	Model 8 Off-axis inclined $a = 12.50$ $\Delta a = 3.4$ $c = 21.1$ $\Delta c = 10.4$ $a/c = 0.59$ $a/t = 0.34$	2.19* * F_1 values	ns 30° 2.33*	ns 30° 2.34*	

Notations:

1. P = axial compressive load
   P_{max} = maximum internal pressure to grow crack
   P_{sf} = stress freezing pressure
2. α = crack depth; $\Delta\alpha$ = crack length growth
   c = half length of crack in fin tip surface; Δc = half crack growth in fin tip surface
3. $\sqrt{Q}$ = approximation of elliptic integral of second kind

$$Q = 1 + 1.464\left(\frac{\alpha}{c}\right)^{1.65} \quad \frac{\alpha}{c} \le 1 \qquad Q = 1.464\left(\frac{\alpha}{c}\right)^{1.65} \quad \frac{\alpha}{c} > 1$$

All flaws were characterized as semi-elliptic flaws of depth α and length 2c.
However, off-axis cracks were neither perfectly semi-elliptic nor planar.

4. ns = near surface slices were used in place of surface slices to avoid a possible surface fringe
5. Slices at $\theta = 60°$ revealed a shear mode in Model 4 but not in Model 8.

Approved for public release; distribution unlimited.

ACKNOWLEDGEMENT

The writers wish to acknowledge the support of Oak Ridge National Laboratory Under Contract No. W7405-Eng-26 and P.O. No. RPO 10230 with ERC Inc. through AFRL at Edwards Air Force Base.

REFERENCES

[1] Smith, C. W., "Experimental Determination of Stress Intensity Factor Distributions in Engineering Problems," *Applied Mechanics Reviews*, Vol. 46, No. 11, Part 2, ,pp. 529-540, November 1993.

[2] Smith, C. W. and Kobayashi, A. S., Chapter 20 of *Handbook on Experimental Mechanics*, A. S. Kobayashi, Ed., pp. 891-956 Prentice-Hall, 1987.

[3] Cotterell, B., "On Brittle Fracture Paths," *International Journal of Fracture Mechanics*, Vol. 1, pp. 96-103, 1965.

[4] Cotterell, B. "Notes on Paths and Stability of Cracks," *International Journal of Fracture Mechanics*, Vol. 2, pp. 526-533, 1966.

[5] Broekhoven, M. J. G., "Computation of Stress Intensity Factors for Nozzle Corner Cracks by Various Finite Element Procedures," Report MMPP 119, Delft University of Technology, Laboratory for Thermal Power and Nuclear Engr. Paper G 416 3rd SMIRT Conference London 1975.

[6] Smith, C. W. and Wiersma, S. J., "Stress Fringe Signatures for Propagating Cracks," *Journal of Engineering Fracture Mechanics*, Vol. 23, No. 1, pp. 229-236, 1986.

[7] Newman, J. C., Jr., Fracture Analysis of Surface and Through Cracks in Cylindrical Pressure Vessels, NASA-TN-D8325, Dec. 1976.

APPENDIX A- Frozen Stress Photoelasticity

When a transparent model is placed in a circularly polarized monochromatic light field and loaded, dark fringes will appear which are proportional to the applied load. These fringes are called stress fringes or isochromatics, and the magnitude of the maximum in-plane shear stress is a constant along a given fringe.

Some transparent materials exhibit mechanical diphase characteristics above a certain temperature, called the critical temperature (T_c). The material, while still perfectly elastic will exhibit a fringe sensitivity of about twenty times the value obtained at room temperature, and its modulus of elasticity will be reduced to about one six-hundredth of its room temperature value. By raising the model temperature above T_c, loading, and then cooling slowly to room temperature, the stress fringes associated with T_c will be retained when the material is returned to room temperature. Since the material is so much more sensitive to fringe generation above T_c than at room temperature, fringe recovery at room temperature upon unloading is negligible. The model may then be sliced without disturbing the "frozen in" fringe pattern and analyzed as a two-dimensional model but containing the three-dimensional effects. In the use of the method to make measurements near crack tips, due to the need to reduce loads above critical temperature to preclude large local deformations, and the use of thin slices, few stress fringes are available by standard procedures. To overcome this obstacle, a refined polariscope is employed to allow the tandem use of the Post and Tardy methods to increase the number of fringes available locally.

In fringe photographs, integral fringes are dark in a dark field and bright in a bright field.

Appendix B

Mode I Algorithm

Beginning with the Griffith-Irwin Equations, we may write, for Mode I, for the homogeneous case,

$$\sigma_{ij} = \frac{K_1}{(2\pi r)^{\frac{1}{2}}} f_{ij}(\theta) + \sigma_{ij}^{\circ} \qquad (i,j = n,z) \tag{1}$$

where:

σ_{ij} are components of stress

Approved for public release; distribution unlimited.

K_1 is SIF
r, θ are measured from crack tip (Fig. B-1)
σ_{ij}° are non-singular stress components
Then, along $\theta = \pi/2$, after truncating σ_{ij}

$$\tau_{nz}^{max} = \frac{K_1}{(8\pi r)^{\frac{1}{2}}} + \tau^{\circ} = \frac{K_{AP}}{(8\pi r)^{\frac{1}{2}}} \quad (2)$$

where:
$\tau^{\circ} = f(\sigma_{ij}^{\circ})$ and is constant over the data range
K_{AP} = apparent SIF
τ_{nz}^{max} = maximum shear stress in nz plane

Normalizing with respect to $\bar{\sigma}$,

$$\therefore \frac{K_{AP}}{\bar{\sigma}(\pi a)^{\frac{1}{2}}} = \frac{K_1}{\bar{\sigma}(\pi a)^{\frac{1}{2}}} + \frac{\sqrt{8}\tau^{\circ}}{\bar{\sigma}}\left(\frac{r}{a}\right)^{\frac{1}{2}} \quad (3)$$

where (Fig. B-1) a = crack length, and $\bar{\sigma}$ = remote normal stress
i.e $\frac{K_{AP}}{\bar{\sigma}(\pi a)^{\frac{1}{2}}}$ vs. $\sqrt{\frac{r}{a}}$ is linear.

From the Stress-Optic Law, $\tau_{nz}^{max} = nf/2t$ where,
n = stress fringe order,
f = material fringe value, and
t = specimen (or slice) thickness
then from Eq. 2

$$K_{AP} = \tau_{nz}^{max}(8\pi r)^{\frac{1}{2}} = \frac{nf}{2t}(8\pi r)^{\frac{1}{2}}$$

where K_{AP} (through a measure of n) and r become the measured quantities from the stress fringe pattern at different points in the pattern.

In part of the present study, instead of normalizing K with respect to $\bar{\sigma}(\pi a)^{1/2}$, we have selected $p_{sf}\sqrt{\pi a/Q}$ as the normalizing factor where $\sqrt{Q}$ is an elliptic integral of the second kind approximated here, as shown in Table I. An example of the determination of F_1 in Table I from test data is given in Fig. B-2.

Mixed Mode Algorithm

The mixed mode algorithm was developed (see Fig. B-3) by requiring that:

$$\lim_{\substack{r_m \to 0 \\ \theta_m \to \theta_m^{\circ}}} \left\{ (8\pi r_m)^{1/2} \frac{\delta(\tau)_{nz}^{max}}{\delta\theta}(K_1, K_2, r_m, \theta_m, \tau_{ij}) \right\} = 0 \quad (4)$$

which leads to:

$$\left(\frac{K_2}{K_1}\right)^2 - \frac{4}{3}\left(\frac{K_2}{K_1}\right) cot\ 2\theta_m^{\circ} - \frac{1}{3} = 0 \quad (5)$$

By measuring θ_m° which is approximately in the direction of the applied load, K_2/K_1 can be determined.

Then writing the stress optic law as:

$$\tau_{nz}^{max} = \frac{fn}{2t} = \frac{K_{AP}^{*}}{(8\pi r)^{\frac{1}{2}}}$$

where K_{ap}^{*} is the mixed mode SIF, one may plot $\frac{K_{AP}^{*}}{\bar{\sigma}(\pi a)^{\frac{1}{2}}}$ vs. $\sqrt{r/a}$ as before, locate a linear zone and extrapolate to $r = 0$ to obtain K^{*}. Knowing, K^{*}, K_2/K_1 and θ_m°, values of K_1 and K_2 may be determined since

$$K^{*} = [(K_1 sin\theta_m^{\circ} + 2K_2 cos\theta_m^{0})^2 + (K_2 sin\theta_m^{\circ})^2]^{\frac{1}{2}} \quad (6)$$

Knowing K^{*} and θ_m°, K_1 & K_2 can be determined from Eqs. 5 and 6. Details are found in Smith and Kobayashi [2].

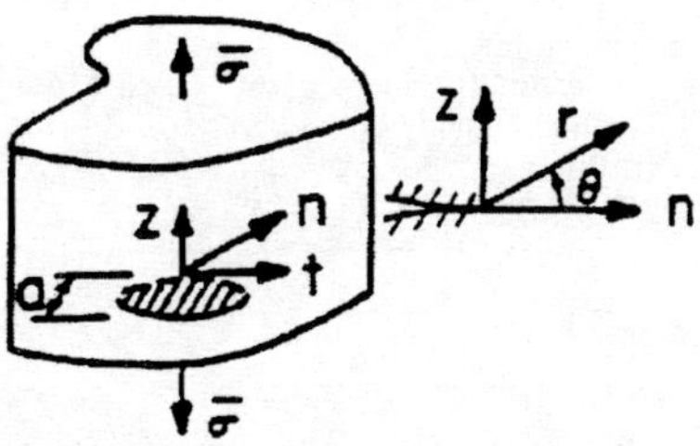

Fig. B-1 Near Tip Notation for Mode I.

Approved for public release; distribution unlimited.

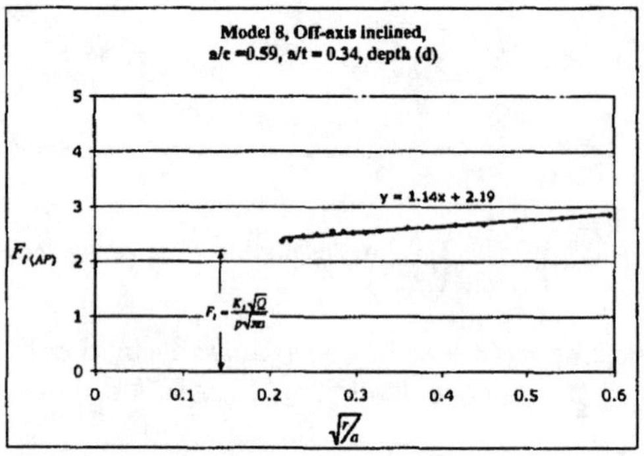

Fig. B-2: Determination of F_1 for Test Data.

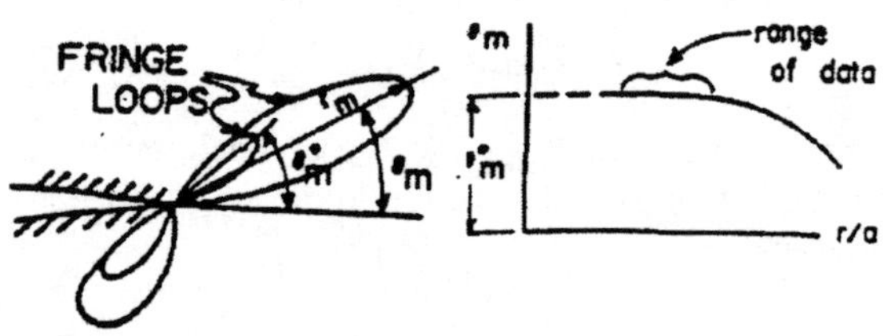

Fig. B-3: Determination of θ_m° for Mixed Mode

Approved for public release; distribution unlimited.

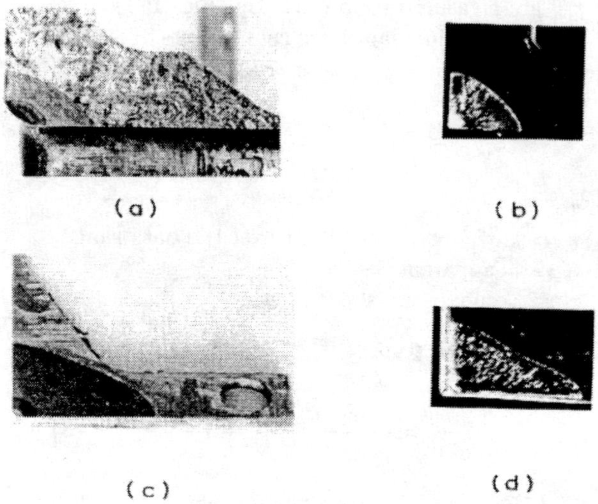

Fig. 1 Comparison of Crack Shapes from Delft Fatigued Steel Model with those grown monotonically in photoelastic model.

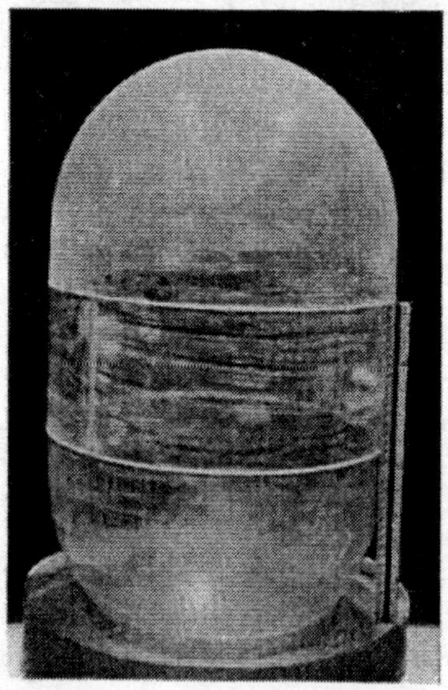

Fig. 2 Photoelastic Model of Nuclear Pressure Vessel Containing Two Diametrically Opposite Nozzles.

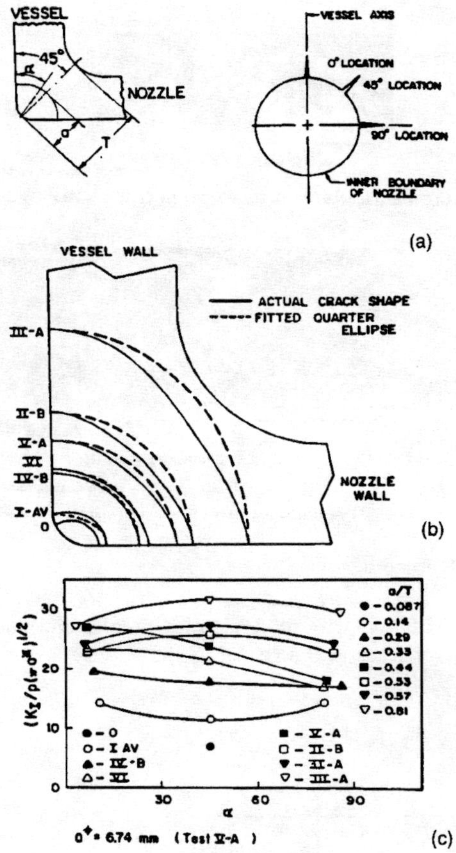

Fig. 3 a) Starter Crack Locations in Nozzle for Thin Walled Pressure Vessel b) Flattening of Crack Shapes with Increasing Depth for $\theta = 0°$ c) Corresponding SIF Distributions p is internal pressure).

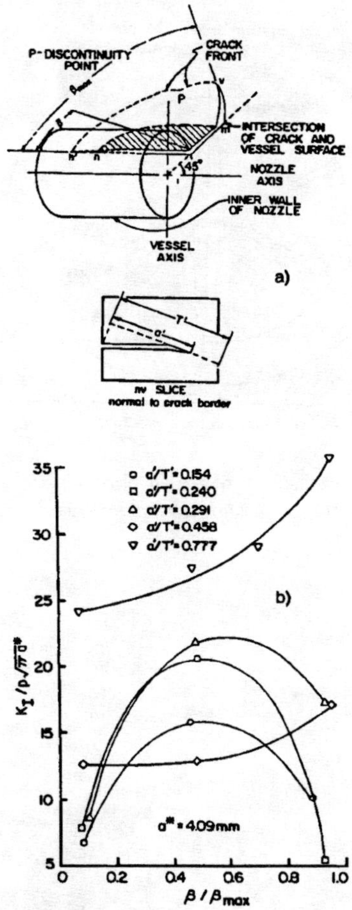

Fig. 4 a) Artist's sketch of planar crack initiated at $\theta = 45°$ b) SIF Distributions for increasing crack depths.

Approved for public release; distribution unlimited.

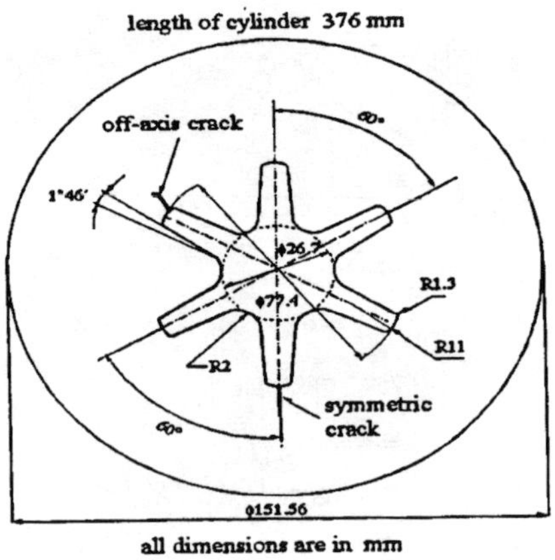

Fig. 5 Cross-section of scaled photoelastic model of motor grain showing critical crack locations.

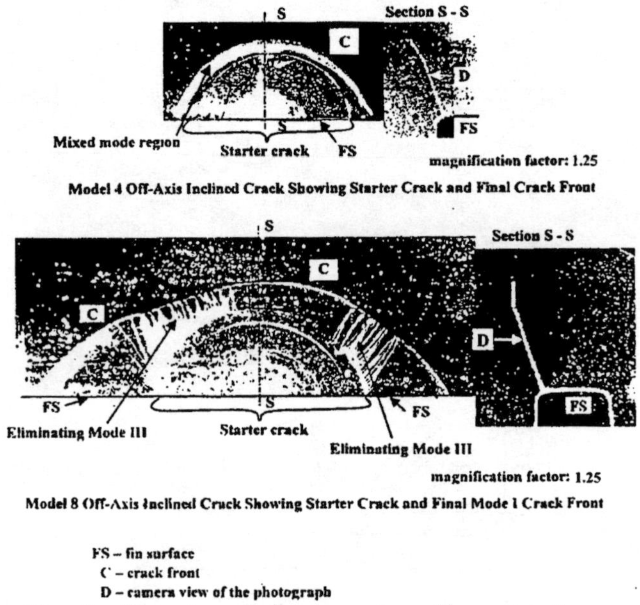

Fig. 6 Plan view of off-axis crack profiles and path of crack at it's center (SS).

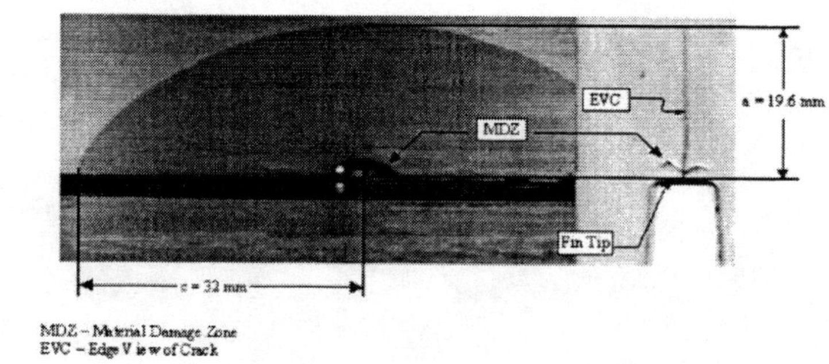

Fig. 7 Plan and crossectional view of a symmetric crack.

Approved for public release; distribution unlimited.

Proceedings of IMECE04
2004 ASME International Mechanical Engineering Congress and Exposition
November 13-20, 2004, Anaheim, California USA

IMECE2004-60834

Investigation of Static Interfacial Fracture in Transparent Glass Fiber Reinforced Polymer Composite Bimaterials Using Photoelasticity

Marius D. Ellingsen
Mechanical and Aerospace Engineering Dept.
University of Missouri-Columbia
Columbia, Missouri, 65211, USA

Sanjeev K. Khanna
Mechanical and Aerospace Engineering Dept.
University of Missouri-Columbia
Columbia, Missouri, 65211, USA

ABSTRACT

The effects of static opening mode loading on orthotropic-orthotropic bimaterial interface cracks and isotropic-orthotropic bimaterial interface cracks has been investigated using the experimental technique of transmission photoelasticity. For successful implementation of this experimental technique transparent glass fiber reinforced polyester composite materials were developed, enabling the direct observation and recording of photoelastic fringes in the orthotropic part of the bi-material. Opening mode stress intensity factors have been calculated using the stress field equations for interfacial cracks in the proposed bimaterials in conjunction with an orthotropic stress-optic law.

Keywords: transparent composites, orthotropic bimaterials, fracture, photoelasticity

INTRODUCTION

Many high-performance structures utilize composite materials for their relative high strength and stiffness to weight ratio. Composite materials have been used in a wide variety of applications for several decades, generally with good results. However, a wide range of possible material anisotropy and a lack of accurate knowledge of the mechanical properties of the fiber-matrix interface both make it difficult to predict the mechanical behavior of composite materials. In particular, it is difficult to predict the mechanical behavior of composite materials containing interfaces and especially cracked interfaces. Examples of interfaces in critical engineering structures are found when considering an airplane whose fuselage is made from composite materials. An example of an interface involving a composite material and a non-composite material would be a cockpit window that is glued onto a surrounding composite material, i.e. the composite airframe. The overlap between the window and the composite is the interface. An example of an interface between two composite materials would be an overlap between two sheets of composite material, such as in composite sheets in the transition between an airplane wing and the airframe proper, or the use of composite patches for repair of composite material based structures.

Fracture of Bimaterial Interfaces in Isotropic/Orthotropic and Orthotropic/Orthotropic Bimaterials

Theoretical work

In recent years the problem of static and dynamic failure of interfaces has received increased attention. There has been much theoretical work done on determining the stress and displacement fields surrounding the tip of a stationary and propagating bimaterial interface crack, as well as determining fracture criteria for such materials. The very first such investigation was on stress fields surrounding the tip of a crack in a bimaterial interface and was performed by Williams in 1959 [1]. Williams proposed a mathematical description of the stress-field surrounding a crack in dissimilar media and found some agreement with physical reality in data provided to him on energy release from fault lines.

There have been several analytical and numerical investigations of that problem and a common issue arises: The occurrence of two sets of solutions, oscillatory and non-oscillatory. An undesirable trait of oscillatory solutions is that they imply that the crack surfaces should wrinkle and overlap near the tips of the crack. In other terms, oscillatory solutions are considered not physically admissible [2]. Williams hypothesized that any pronounced oscillatory character of the stress would be confined to a region very close to the crack tip. For investigation of bimaterial interfaces it is not satis factory to let the matter rest at that; the near-tip stress fields are of great importance for the problem of crack propagation. According to Gdoutos and Papakaliatakis [3] an oscillatory stress state occurs in isotropic materials when traction-free crack surfaces are used as a boundary condition, which is not always realistic. In 1989 Qu and Bassani [4] derived and presented a necessary and sufficient condition for no oscillations in the crack tip fields of bicrystal interfaces. In brief, if the in-plane and anti-plane deformations of each crystal decouple in the interface coordinate system then the interface crack has the standard square-root singularity. This does not always occur. Bassani and Qu [5] further showed that explicit solutions for the stress fields are obtained when the conditions for non-oscillatory stress fields are met. For numerical analyses Cho, et al. [6]

proposed an extrapolation method that avoided the problem of oscillatory near-tip stress fields.

For analytical methods, Deng [7 8] presented complex series expansions for growing cracks in isotropic bimaterials, and later also proposed general crack tip fields for both stationary and growing cracks in anisotropic bimaterials. These stress fields also contain both oscillatory and non-oscillatory solutions. Deng states that non-oscillatory solutions *can* occur, in homogeneous anisotropic bimaterials.

Experimental work

The theoretical work has been augmented by experimental research on static [9-12] and dynamic fracture [13-23] of bimaterial interfaces. Of particular interest is research performed using the method of photoelasticity. Photoelasticity has long been used for qualitative and quantitative analysis of structures, and has been applied to the problem of finding stress intensity factors in transparent materials

Dally and Sanford performed early experimental work demonstrating the use of photoelasticity for static and also dynamic fracture studies [9-11]. Early photoelastic investigations of fracture were performed on isotropic homogeneous materials, later photoelasticity was used to investigate the static mechanical behavior of non-isotropic materials [12].

A photoelastic investigation of bimaterial interfaces was first performed on bimaterials where both half-planes where isotropic, i.e. isotropic/isotropic bimaterials; Gdoutos and Papakaliatakis [13] were the first to perform such a study on the static behavior of an interface crack in an isotropic/isotropic bimaterial. They demonstrated that the photoelastic method was applicable to this problem and presented the mathematical and experimental methods to perform such an analysis. Lu and Chiang [14] expanded on this work and proposed two different approaches to determining the stress-intensity factors from the isochromatic fringe patterns.

More recently photoelastic analysis has been applied to isotropic/non-isotropic bimaterials. What has predominantly been done to explore the behavior of non-isotropic bimaterials has been to create a bimaterial with one isotropic, transparent half-plane and another non-isotropic non-transparent half-plane. The mechanical behavior of the non-isotropic half-plane has been inferred from what could be observed in the transparent isotropic half-plane.

One aspect of this field of research that has not been explored is experimental work on the static and dynamic behavior of interfaces in wholly non-isotropic bimaterials, i.e. the behavior of bi-materials where both half-planes are non-isotropic. This has been due to difficulty in creating non-isotropic materials that have sufficient transparency and birefringent characteristics for photoelastic analysis. Such a material has now been developed by this research group and it has been successfully applied in the study of Mode-I stress intensity factors in orthotropic composite materials [23]. Thus, it is now feasible to undertake experimental photoelastic studies of the static and dynamic behavior of interfaces in wholly non-isotropic bimaterials.

Modeling interfaces

For modeling purposes, an interface is created by joining two bi-materials together with a butt joint. Normally, composite materials are not joined with butt joints, but the butt joint specimens serve the purpose well for investigating the mechanical behavior of both butt and lap joints, as will be explained in the next few paragraphs:

Butt-joint bi-material specimens are made in two general configurations, for pure tension (Figure 1) and combined loading (Figure 2).

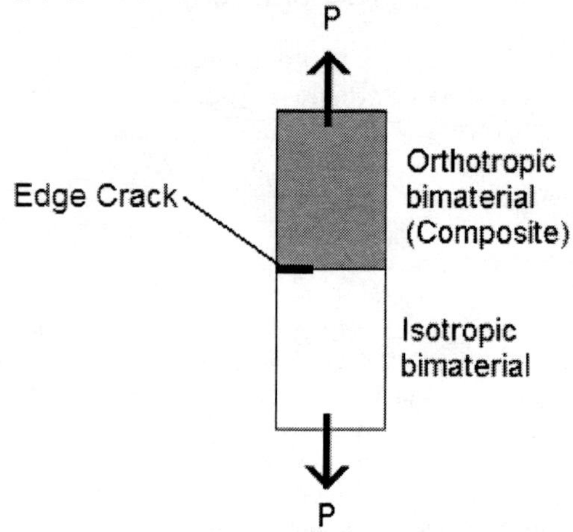

Figure 1 Pure Tension Specimen

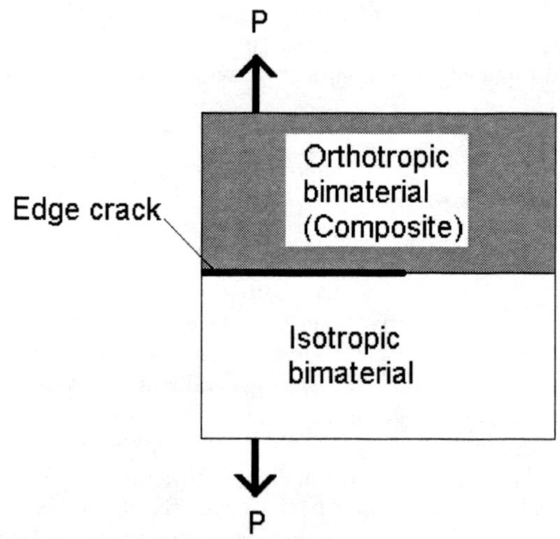

Figure 2 Combined Loading Specimen

The current investigation focuses on pure tension specimens with an edge crack in the interface between the two bi-materials. Load is applied to the geometrical centerline of the specimens and the mechanical behavior is investigated. Since loading occurs along the centerline these specimens are under a pure opening mode (Mode I) loading condition. Combined loading specimens will be the subject of a later investigation.

Combined loading specimens have an edge crack in the interface between the two bi-materials, as illustrated in Figure 2, and the load is applied away from the geometrical center of the specimen. This is also initially an opening mode loading condition. The angle of the interface and the crack between the two bi-materials will then be varied, and a mixed-mode loading condition will occur, with modes I and II present. A mixed-mode loading situation in a butt joint is very similar to the loading conditions that are expected to occur in lap joints under static and dynamic loads. Thus, butt-joined specimens are expected to give results that also are valid for lap-joined materials.

Analytic Background

The methodology used is as follows: first the region around the crack tip is used to obtain the fringe orders at many points from the isochromatic fringe patterns surrounding the crack tip, and then use the stress-field equations in conjunction with the appropriate stress-optic law to determine the stress field. Once the stress field has been determined from the optical data, the results can be fed back into the stress-optic law in order to generate a fringe pattern. If the generated fringe pattern qualitatively and quantitatively matches the observed fringe pattern, then the solution is verified and the proposed stress-field equations have been experimentally verified [24]. However, we have not reported fringe generation results in this paper.

Lee, [15], developed static and dynamic stress-field equations for both isotropic/orthotropic and orthotropic/orthotropic bi-materials. These are presented below and can also be used for static cases by using a crack velocity equal to zero. Using the stress-optic laws it is possible to relate the observed fringe patterns to the state of stress in the material. In the following treatment, it is assumed that the specimens remain in a plane stress state. This may not hold true for the entire region of the specimens, a region very close to the crack tip may be under a state of plane strain, and so no data is collected from this region. Furthermore, far away from the crack tip the stress state becomes dominated by the remotely applied stress and therefore a maximum radius is specified, beyond which no data is collected. Thus, an annular region is specified from which data is collected, with an inner diameter equal to half the specimen thickness, and an outer diameter equal to twice the specimen thickness.

The asymptotic stress fields for a steadily propagating crack in an isotropic bi-material, represented as oscillatory stress fields with odd power series are given by Lee [15]. For an orthotropic bi-material, the oscillatory stress fields having an odd power series (n = 1,3,5,...) are as follows [15]:

$$
\begin{aligned}
\sigma_{xn} = \sum_{n=odd}^{\infty} \frac{K_n^{\;0}}{2\sqrt{2\pi}\,D\cosh(\varepsilon\pi)} & \left[\begin{aligned} &\left(M_b - p^2\right)\left\{ e^{\varepsilon(\pi-\theta_l)}\overline{A}\cos\left(\varepsilon\ln r_l + \frac{n-2}{2}\theta_l\right) + e^{-\varepsilon(\pi-\theta_l)}A\cos\left(\varepsilon\ln r_l + \frac{n-2}{2}\theta_l\right)\right\} r_l^{\left(\frac{n-2}{2}\right)} \\ &-\left(M_b - q^2\right)\left\{ e^{\varepsilon(\pi-\theta_s)}\overline{B}\cos\left(\varepsilon\ln r_s + \frac{n-2}{2}\theta_s\right) + e^{-\varepsilon(\pi-\theta_s)}B\cos\left(\varepsilon\ln r_s + \frac{n-2}{2}\theta_s\right) r_s^{\left(\frac{n-2}{2}\right)}\right\} \end{aligned} \right] \\
+ \sum_{n=odd}^{\infty} \frac{K_n^{\;*}}{2\sqrt{2\pi}\,D\cosh(\varepsilon\pi)} & \left[\begin{aligned} &-\left(M_b - p^2\right)\left\{ e^{\varepsilon(\pi-\theta_l)}\overline{A}\sin\left(\varepsilon\ln r_l + \frac{n-2}{2}\theta_l\right) + e^{-\varepsilon(\pi-\theta_l)}A\sin\left(\varepsilon\ln r_l + \frac{n-2}{2}\theta_l\right)\right\} r_l^{\left(\frac{n-2}{2}\right)} \\ &+\left(M_b - q^2\right)\left\{ e^{\varepsilon(\pi-\theta_s)}\overline{B}\sin\left(\varepsilon\ln r_s + \frac{n-2}{2}\theta_s\right) + e^{-\varepsilon(\pi-\theta_s)}B\sin\left(\varepsilon\ln r_s + \frac{n-2}{2}\theta_s\right) r_s^{\left(\frac{n-2}{2}\right)}\right\} \end{aligned} \right]
\end{aligned}
$$

$$
\begin{aligned}
\sigma_{yn} = \sum_{n=odd}^{\infty} \frac{K_n^{\;0}}{2\sqrt{2\pi}\,D\cosh(\varepsilon\pi)} & \left[\begin{aligned} &\left(1 + M_a\right)\left\{ e^{\varepsilon(\pi-\theta_l)}\overline{A}\cos\left(\varepsilon\ln r_l + \frac{n-2}{2}\theta_l\right) + e^{-\varepsilon(\pi-\theta_l)}A\cos\left(\varepsilon\ln r_l + \frac{n-2}{2}\theta_l\right)\right\} r_l^{\left(\frac{n-2}{2}\right)} \\ &-\left(1 + M_a\right)\left\{ e^{\varepsilon(\pi-\theta_s)}\overline{B}\cos\left(\varepsilon\ln r_s + \frac{n-2}{2}\theta_s\right) + e^{-\varepsilon(\pi-\theta_s)}B\cos\left(\varepsilon\ln r_s + \frac{n-2}{2}\theta_s\right) r_s^{\left(\frac{n-2}{2}\right)}\right\} \end{aligned} \right] \\
+ \sum_{n=odd}^{\infty} \frac{K_n^{\;*}}{2\sqrt{2\pi}\,D\cosh(\varepsilon\pi)} & \left[\begin{aligned} &-\left(1 + M_a\right)\left\{ e^{\varepsilon(\pi-\theta_l)}\overline{A}\sin\left(\varepsilon\ln r_l + \frac{n-2}{2}\theta_l\right) + e^{-\varepsilon(\pi-\theta_l)}A\sin\left(\varepsilon\ln r_l + \frac{n-2}{2}\theta_l\right)\right\} r_l^{\left(\frac{n-2}{2}\right)} \\ &+\left(1 + M_b\right)\left\{ e^{\varepsilon(\pi-\theta_s)}\overline{B}\sin\left(\varepsilon\ln r_s + \frac{n-2}{2}\theta_s\right) + e^{-\varepsilon(\pi-\theta_s)}B\sin\left(\varepsilon\ln r_s + \frac{n-2}{2}\theta_s\right) r_s^{\left(\frac{n-2}{2}\right)}\right\} \end{aligned} \right]
\end{aligned} \tag{1}
$$

$$
\begin{aligned}
\tau_{xyn} = \sum_{n=odd}^{\infty} \frac{K_n^{\;0}}{2\sqrt{2\pi}\,D\cosh(\varepsilon\pi)} & \left[\begin{aligned} &\alpha_l\left\{ e^{\varepsilon(\pi-\theta_l)}\overline{A}\sin\left(\varepsilon\ln r_l + \frac{n-2}{2}\theta_l\right) - e^{-\varepsilon(\pi-\theta_l)}A\sin\left(\varepsilon\ln r_l + \frac{n-2}{2}\theta_l\right)\right\} r_l^{\left(\frac{n-2}{2}\right)} \\ &+\alpha_s\left\{ -e^{\varepsilon(\pi-\theta_s)}\overline{B}\sin\left(\varepsilon\ln r_s + \frac{n-2}{2}\theta_s\right) + e^{-\varepsilon(\pi-\theta_s)}B\sin\left(\varepsilon\ln r_s + \frac{n-2}{2}\theta_s\right) r_s^{\left(\frac{n-2}{2}\right)}\right\} \end{aligned} \right] \\
+ \sum_{n=odd}^{\infty} \frac{K_n^{\;*}}{2\sqrt{2\pi}\,D\cosh(\varepsilon\pi)} & \left[\begin{aligned} &\alpha_l\left\{ e^{\varepsilon(\pi-\theta_l)}\overline{A}\cos\left(\varepsilon\ln r_l + \frac{n-2}{2}\theta_l\right) - e^{-\varepsilon(\pi-\theta_l)}A\cos\left(\varepsilon\ln r_l + \frac{n-2}{2}\theta_l\right)\right\} r_l^{\left(\frac{n-2}{2}\right)} \\ &+\alpha_s\left\{ -e^{\varepsilon(\pi-\theta_s)}\overline{B}\cos\left(\varepsilon\ln r_s + \frac{n-2}{2}\theta_s\right) + e^{-\varepsilon(\pi-\theta_s)}B\sin\left(\varepsilon\ln r_s + \frac{n-2}{2}\theta_s\right) r_s^{\left(\frac{n-2}{2}\right)}\right\} \end{aligned} \right]
\end{aligned}
$$

where

$$A=\alpha_s+(1+M_a)\eta \qquad \overline{A}=\alpha_s-(1+M_a)\eta$$

$$B=\alpha_l+(1+M_a)\eta \qquad \overline{B}=\alpha_l-(1+M_a)\eta$$

$$B_{12}=\frac{1}{2}\left[2a_{12}+a_{66}+\rho c^2\left(a_{12}^{\ 2}-a_{11}a_{66}-a_{11}a_{22}\right)\right]\frac{1}{a_{11}}$$

$$D_k=\left[(1+M_a)(\alpha_s-\alpha_l)\right]_k$$

$$h_{11}=(l_{11})_1-(l_{11})_2$$

$$h_{12}=(l_{12})_1+(l_{12})_2$$

$$h_{21}=(l_{21})_1+(l_{21})_2$$

$$K_{66}=\frac{1}{a_{11}}\left\{a_{22}+\rho c^2\left[a_{12}^{\ 2}-a_{22}a_{66}-a_{11}a_{22}+\rho c^2a_{66}\left(a_{11}a_{22}-a_{12}^{\ 2}\right)\right]\right\}$$

$$(l_{11})_k=\left\{\frac{p_s\alpha_l-p_l\alpha_s}{D}\right\}_k=\left\{\frac{q_s-q_l}{\alpha_s-\alpha_l}\right\}_k$$

$$(l_{12})_k=\left\{\frac{(1+M_\alpha)(p_l-p_s)}{D}\right\}_k$$

$$(l_{21})_k=\left\{\frac{\alpha_sq_l-\alpha_lq_s}{D}\right\}_k$$

$$M_a=\rho c^2(a_{12}-a_{11}) \qquad M_b=\rho c^2(a_{12}-a_{22})$$

$$p=\sqrt{B_{12}-\sqrt{B_{12}^{\ 2}-K_{66}}}$$

$$q=\sqrt{B_{12}+\sqrt{B_{12}^{\ 2}-K_{66}}}$$

$$r_l=r\sqrt{\cos^2\theta+p^2\sin^2\theta}$$

$$r_s=r\sqrt{\cos^2\theta+q^2\sin^2\theta}$$

$$Z_l=p \quad Z_s=q$$

$$\alpha_l=p+a_{22}\frac{\rho c^2}{p}-p\rho c^2a_{11}-\frac{(\rho c^2)^2}{p}\left(a_{11}a_{22}-a_{12}^{\ 2}\right)$$

$$\alpha_s=q+a_{22}\frac{\rho c^2}{q}-q\rho c^2a_{11}-\frac{(\rho c^2)^2}{q}\left(a_{11}a_{22}-a_{12}^{\ 2}\right)$$

$$\beta=\frac{h_{11}}{\sqrt{h_{12}h_{21}}}$$

$$\varepsilon=\frac{1}{2\pi}\ln\frac{1-\beta}{1+\beta}$$

$$\theta_l=\arctan(Z_l\tan\theta) \qquad \theta_s=\arctan(Z_s\tan\theta)$$

In the odd power series, if n is chosen equal to 1, then $K_n^{\ 0}$ and $K_n^{\ *}$ become K_I and K_{II}, respectively.

The non-oscillatory stress fields (with even power series) are also given by Lee [15]

Stress-optic law

The observed fringe patterns are related to the mechanical stresses. Hyer and Liu [16] presented the following stress-optic law for orthotropic materials, which was shown to be applicable for dynamic cases as well by Haixiao, et al. [17]:

$$N_T=\left\{\left(\frac{\sigma_{xx}}{C_1}-\frac{\sigma_{yy}}{C_2}+\frac{\tau_{xy}}{C_3}+N_R\cos 2\delta_R\right)^2+\left[2\left(\frac{\sigma_{xx}-\sigma_{yy}}{C_4}+\frac{\tau_{xy}}{C_5}\right)+N_R\sin 2\delta_R\right]^2\right\}^{\frac{1}{2}} \quad (2)$$

where

$$\frac{1}{C_1}=d\left(\frac{\cos^2\theta}{f_1}-\frac{\sin^2\theta}{f_2}\right)$$

$$\frac{1}{C_2}=d\left(\frac{\cos^2\theta}{f_2}-\frac{\sin^2\theta}{f_1}\right)$$

$$\frac{1}{C_3}=d\left(\frac{1}{f_1}+\frac{1}{f_2}\right)\sin 2\theta$$

$$\frac{1}{C_4}=-\frac{d\sin\theta\cos\theta}{f_{12}}$$

$$\frac{1}{C_5}=\frac{d\left(\cos^2\theta-\sin^2\theta\right)}{f_{12}}$$

N_T is the observed fringe order, N_R is the residual fringe order (if residual fringes are present), d_R is the residual isoclinic angle in the material coordinate system, f_1, f_2, and f_{12} are the dynamic stress-fringe values of the material, ? represents the fiber angle in reference to the x-axis and d is the thickness of the specimen. The stress-optic law is applied to static cases by substituting static stress-fringe values for the dynamic stress-fringe values. For isotropic materials, the inverse C_i terms are simplified as the stress-fringe values are equal for all material directions, i.e. f_1, f_2, and f_{12} are all equal to each other.

Overdeterministic Analysis:

Static fringe patterns are recorded with a digital imaging system. The fringe patterns are then examined with image analysis software and a number of data points are taken, where each data point consists of the fringe order N at that point, and the location (r, ?) of the point in the material coordinate system. If the *exact* stresses and fringe orders at each point were known then both sides of the stress-optic law would equal each other. However in practice this is not the case so the *difference* between the two sides has been defined as an error function g, written as

$$g=\left\{\begin{aligned}&\left[\left(\frac{\sigma_{xx}}{C_1}-\frac{\sigma_{yy}}{C_2}+\frac{\tau_{xy}}{C_3}+N_R\cos 2\delta_R\right)^2\right.\\&\left.+\left[2\left(\frac{\sigma_{xx}-\sigma_{yy}}{C_4}+\frac{\tau_{xy}}{C_5}\right)+N_R\sin 2\delta_R\right]^2\right]\end{aligned}\right\}^{\frac{1}{2}}-N_T \quad (3)$$

There will be one equation for each data point, containing data on fringe location and fringe order. As expressed above, the error function g determines the difference between the observed fringe order at each point, and what the calculated fringe order at that point should be. The calculated fringe order is determined on basis of the calculated stresses at that point and the coordinates of the point. The system is said to be overdeterministic as the number of unknowns is less than the number of simultaneous equations. An iterative approach is used to solve the system of equations, where g should decrease for each iteration. Once g has come sufficiently close to zero the calculated and observed fringe orders are virtually equal. The calculated fringe order contains the calculated stress fields, and so the stress fields can be extracted from the above equations.

The overdeterministic method was implemented in a mathematical software program called Maple, which works well for implementing systems of very large equations. Typically 30 data points are used with 25 iterations. With a two-term odd power series the iterative calculations take 5-10 minutes to complete on a Pentium 4 computer system. Though for common calculations 20-25 iterations are completed in a moment on a modern computer system, the equations in question are very large and so each iteration involves a large amount of calculation.

Experimental Procedures

The composite material used in this investigation was a woven E-glass cloth reinforced polyester matrix composite. The matrix used was Polyester resin. The polymer was cured at room temperature for two days and also chemically tailored to match the refractive index of the glass fibers with the polymer. The reinforcement was seven layers of bi-directionally woven glass cloth giving a fiber volume fraction of about 18%. The thickness of the composite material was 6.35 mm (1/4 inch); it was fabricated by a hand lay-up technique in 330 mm by 510 mm (13 by 20 inches) sheets. A 6.35 mm (1/4 inch) thick aluminum frame was laid on top of an acrylic plate clad with a Mylar sheet, which also acted as a spacer. Resin was poured into the frame, and a 330 mm by 510 mm (13 by 20 inches) fiberglass cloth was laid on top. A spatula was used to purge any air bubbles that may have been trapped beneath the cloth. Resin was then poured on top of the cloth, and a new layer of cloth was laid down. This process was repeated until all layers of cloth were laid down. Another Mylar-clad acrylic plate was laid on top of the frame, the top and bottom plates were clamped with C-clamps firmly against the aluminum frame. This arrangement was then erected sideways. The upright position causes any lack of resin to form an edge void rather than a large surface void. Transparency was achieved by chemically matching the refractive indices of the polyester matrix and glass fiber reinforcement.

Photoelastic Properties

The photoelastic response of a material is a measure of which fringe order will be observed in, for instance, the dark field of the polariscope for a given stress in the material. The photoelastic response can be found by loading tensile specimens in the polariscope and recording fringe orders vs. load. For this purpose, strips were cut from the fabricated composite plates with reinforcement directions of 0°/90° and ±45°. The nominal thickness of the specimens was 6.35 mm (1/4 inch).

The composite strips were loaded in tension in the "dark field" of the polariscope. In the dark field, a specimen under no load appears dark. This is the zero-order fringe. When tensile load is applied the specimen gradually appears brighter and then darker again as the load increases. The load was recorded each time the specimen appeared "dark", and the data was plotted as fringe order vs. calculated normal stress. The specimens were loaded until failure. The photoelastic response f_σ was calculated using the expression

$$f=\frac{\sigma}{N}\cdot t=\frac{P\cdot t}{N\cdot w\cdot t}=\frac{P}{N\cdot w} \quad (4)$$

where σ is the normal stress, P is the applied load, w the specimen width, t the thickness, and N the observed finge order. A linear least squares curve fit was used across the data range yielding

$$f_L=f_T=58\frac{kPa\cdot m}{fringe} \quad \text{and} \quad f_{LT}=28\frac{kPa\cdot m}{fringe}$$

f_L and f_T are the photoelastic responses of material samples cut along the 0° and 90° reinforcement directions, respectively, while f_{LT} is the photoelastic response of the material samples cut in the ±45° reinforcement directions. The mechanical properties of the composite are equal in the L and T directions, so it is reasonable to expect that f_L equals f_T. This was also found to be the case. The photoelastic response was found to be linear with respect to load throughout the loading range.

Testing

During all tests a digital camera was used to record fringe patterns. A transparent plastic ruler was attached to the edge of the specimen with modeling clay. The ruler has a grid on it which provided a means of identifying dimensions in the photographs, as well as making it easier to focus the digital camera. The photographs were of high resolution and this made it possible to greatly enlarge the image area in question while retaining a clear image. This also facilitated determining the location of the crack tip; the location could be clearly observed and marked. The image analysis program had a feature that gave coordinates for each "pixel" in the image. Thus, the location of the crack tip could be accurately determined. Once the location of the crack tip was established, an annular region was drawn around the crack tip in the photograph, from which data points were taken. The inner diameter of the region was equal to the sample thickness. Very close to the crack tip the material is under a three-dimensional stress state rather than plane stress, so a plane stress solution would not be valid inside this region. Far away from the crack tip the stress field is dominated by remote stresses rather than the effect of the crack tip, so the validity of the mathematical solution decreases at a larger distance from the crack tip. Hence the outer diameter of the annular region was chosen to be four times the thickness of the material. Each data point represented a position relative to the crack tip, and the fringe order at that point. Errors may be made in determining the exact location of the crack tip, and this would affect the accuracy of the data taken in that the data points would have contain inaccurate information regarding their position relative to the actual crack tip position. However, with the current approach any error in the determination of the crack tip location is thought to be small, as are errors in stress intensity calculations that would result from such inaccuracies.

Three bimaterial specimen types were tested: composite-composite, composite-polyester resin, and composite-aluminum. The former is an orthotropic-orthotropic bimaterial while the two latter are isotropic-orthotropic bimaterials. All specimens were made to the same dimensions, namely thickness t = 6.4 mm (1/4 inch), width w = 76 mm (3 inches) and overall length 305 mm (12 inches). The specimens are held by bolts that are inserted into 19 mm (3/4 inch) holes which have been machined in each end of the specimens.

For testing a Terco MT3100 polariscope was used along with an in-house designed loading frame. This loading frame employs a hydraulic cylinder with a hand pump to apply load. A digital imaging system with a Pulnix TM1040 camera and Epix imaging software was used to record the fringe patterns. Each specimen was gradually loaded to failure while recording load and corresponding fringe patterns.

Fringe Multiplication

In some cases the specimens tended to break before higher order fringes became visible around the interface crack tip, thus yielding only a well developed first order or fractional order fringe, as seen in figure 3 below. To increase the amount of information that could be extracted from the photoelastic fringe patterns a fringe multiplication procedure developed by Chen [27] was implemented.

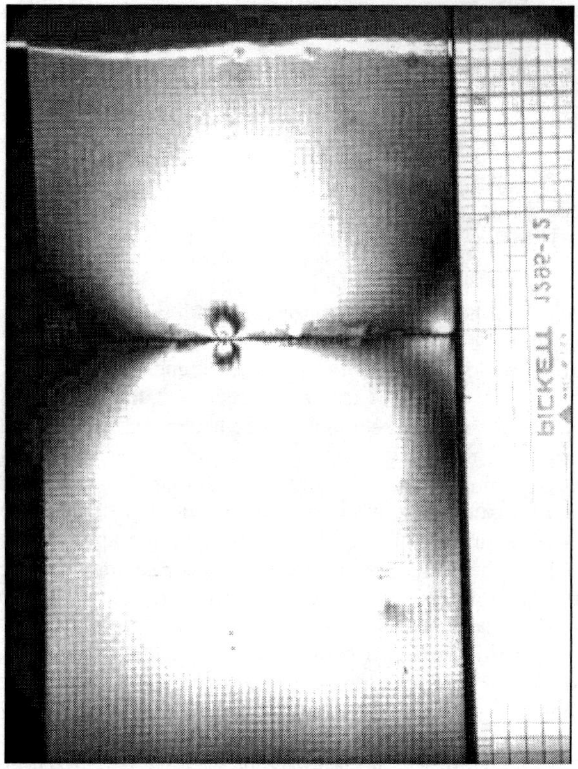

Figure 3 – Full view of composite-composite bimaterial specimen under load.

The fringe mu ltiplication method is readily implementable in a mathematics program such as MathCad. In essence, two pictures are used to multiply the fringe patterns, a picture of a loaded specimen and a picture of a specimen under no load. The intensities are then compared, that is, the intensity value of each pixel in the picture showing the loaded specimen is compared to the intensity value for the corresponding pixel in the picture showing the unloaded specimen. A relative retardation is then computed by

$$\Delta = 2\cos^{-1}\left(\frac{I_2}{I_1}\right)^{\frac{1}{2}} \tag{7}$$

Where ? is the relative retardation and I_1 and I_2 are the intensities in the unloaded and loaded photoelastic model, respectively. A new set of intensities (and thus a new fringe pattern) can then be calculated by

$$I_p = \cos^2(\Delta) \tag{8}$$

Equation 7 will double the fringe patterns, whereas the following equation can be used to triple the fringe pattern:

$$I_p = \cos^2\left(\frac{3\Delta}{2}\right) \tag{9}$$

These equations can be used successively and in conjunction with each other to further multiply the fringe patterns. The method is demonstrated in two different specimens below. The first set of images depicts a composite disc in uniaxial compression. The disc has been compressed with a C-clamp with low force in order to give few fringes, as can be seen in Figure 4. Figure 5 shows a fringe pattern that has been multiplied threefold, and the fringes in Figure 6 have been multiplied sixfold. The second set of images shows a close-up of a fringe pattern around the tip of a crack in the interface of a composite-composite bimaterial. Figure 7 shows the recorded fringe pattern, while figures 8 and 9 show that fringe pattern multiplied threefold and sixfold, respectively.

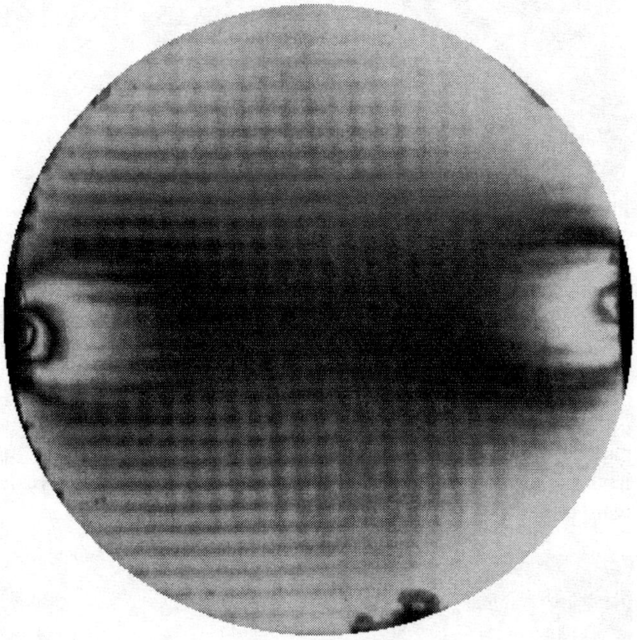

Figure 4 – Circular composite disc in compression

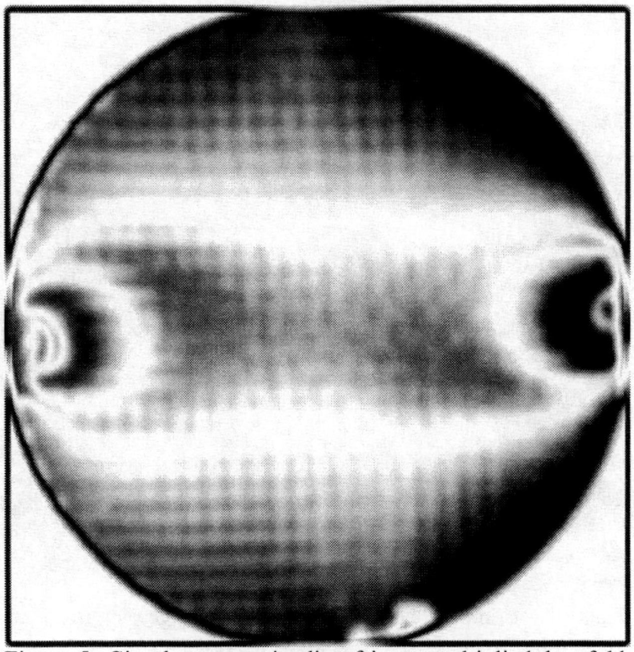

Figure 5– Circular composite disc, fringes multiplied threefold

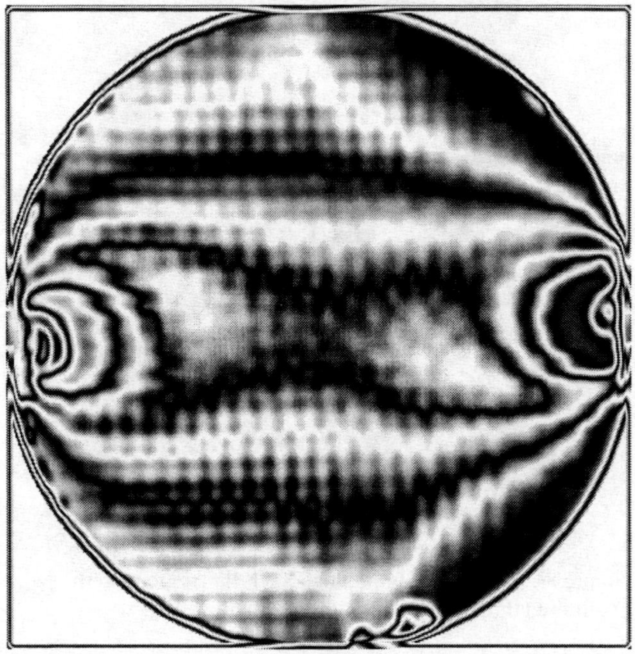

Figure 6 – Circular composite disc, fringes multiplied sixfold

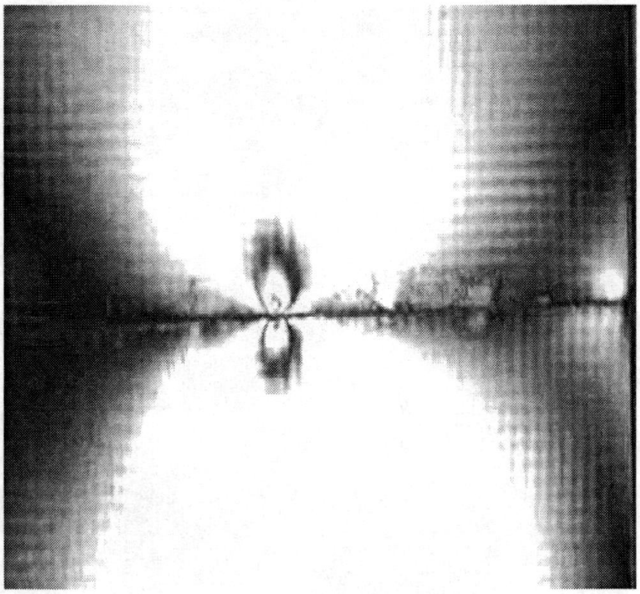

Figure 7 – Composite bimaterial crack tip fringe pattern – close-up.

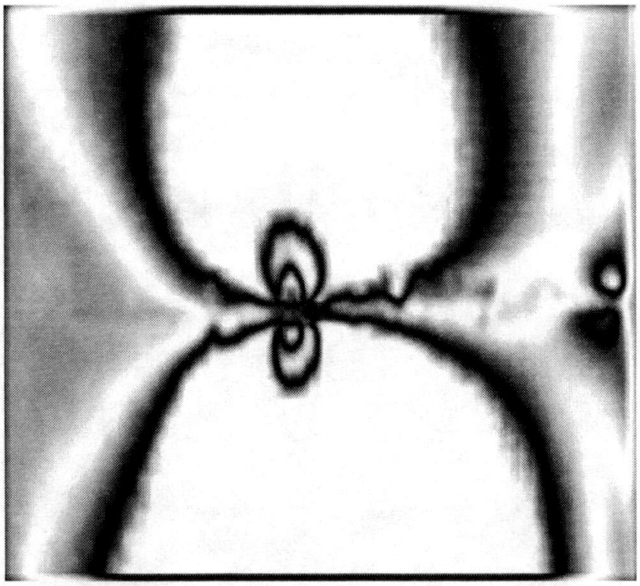

Figure 8 – Composite bimaterial crack tip fringe pattern multiplied threefold

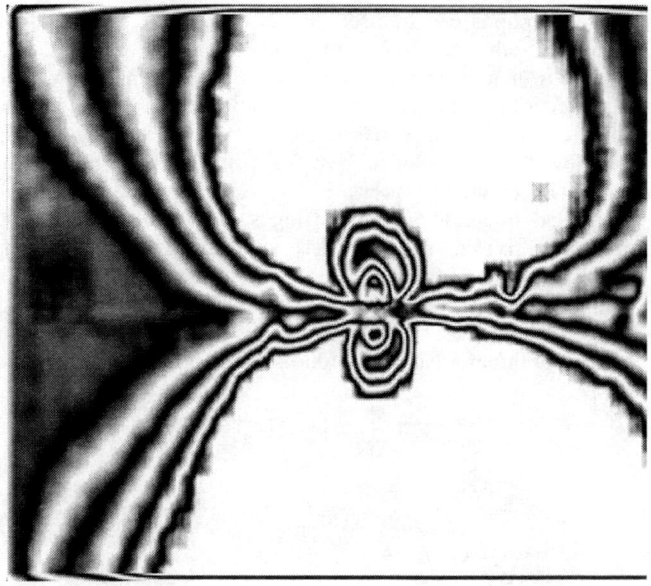

Figure 9 - Composite bimaterial crack tip fringe pattern multiplied sixfold

7. Results and Discussion:

Close-up images of the fringe patterns from the three specimen types are shown below. Figure 10 shows the fringe pattern obtained for a composite-composite bimaterial, i.e. an orthotropic-orthotropic bimaterial. Figure 11 shows the fringe pattern obtained for a composite-aluminum bimaterial, i.e. an isotropic-orthotropic bimaterial, as is the composite-polyester resin bimaterial which is shown in Figure 12.

Figure 10 Close-up of fringe pattern surrounding crack tip in composite-composite bimaterial

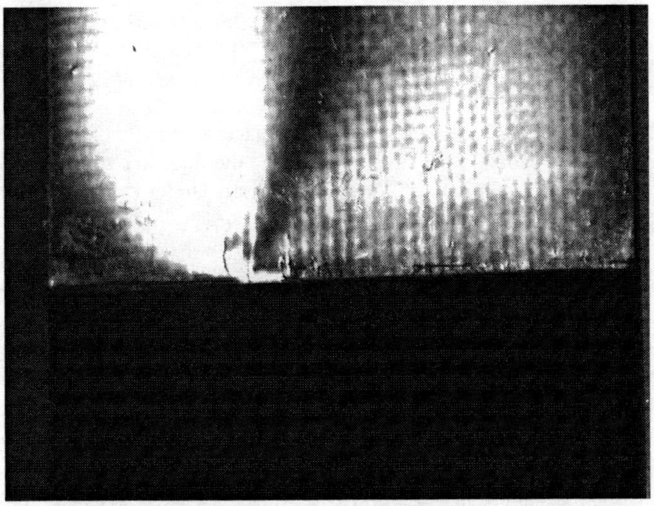

Figure 11a Close-up of fringe pattern surrounding crack tip in composite-aluminum bimaterial

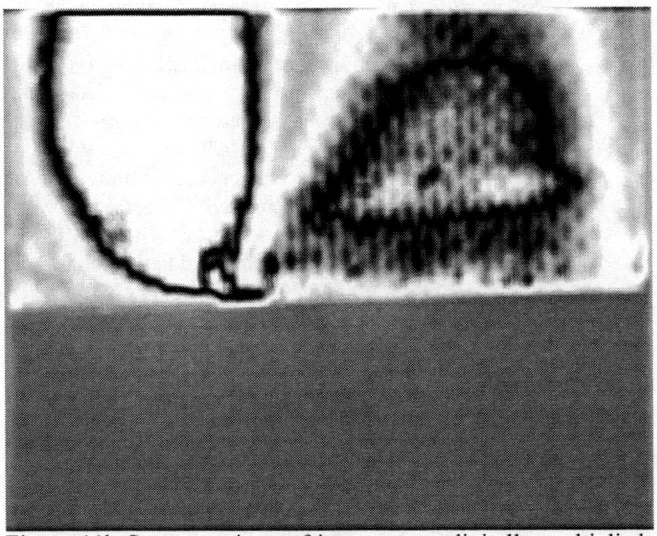

Figure 11b Same specimen, fringe pattern digitally multiplied threefold.

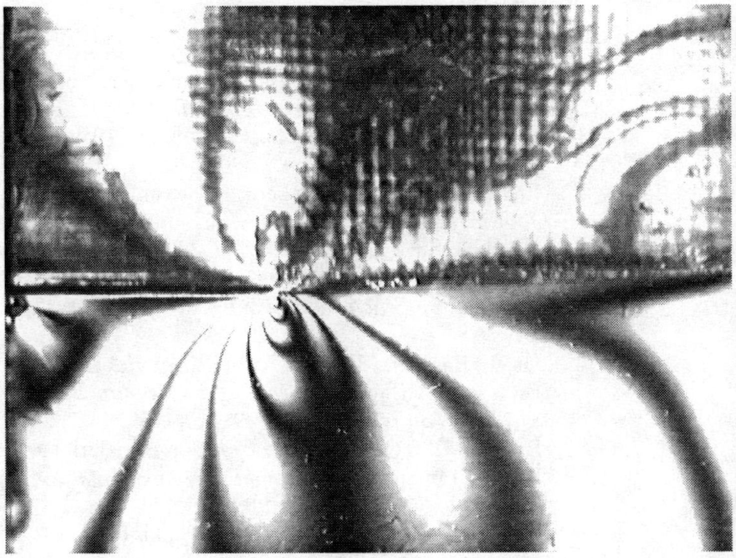

Figure 12 Close-up of fringe pattern surrounding crack tip in composite-polyester resin bimaterial

The K_I and K_{II} values were calculated based on the methodology outlined above, using a two-term odd power series, that is n=1 . The calculations yielded the following result:

(1) Composite-composite interface:

load P = 0.77 kN (172 lb), a/w = 0.45

K_I = 0.73 MPa$\sqrt{m}$
K_{II}= 0.09 MPa$\sqrt{m}$

(2) Composite-aluminum interface:

load P = 1.0 kN (227 lb), a/w = 0.33

K_I = 0.65 MPa$\sqrt{m}$
K_{II}= 0.16 MPa$\sqrt{m}$

(3) Composite-polyester interface:

load P = 2.63 kN (592 lb), a/w = 0.40

K_I = 1.59 MPa$\sqrt{m}$
K_{II}= 0.16 MPa$\sqrt{m}$

SUMMARY

In this investigation transparent glass fiber reinforced polymer composites have been used to investigate interface fracture in orthotropic-orthotropic bimaterial interfaces. In some cases it may be necessary to use fringe multiplication methods for analyzing isochromatic fringe patterns around the crack tip. In addition the crack growth and damage mechanisms can be recorded in real time.

REFERENCES

1. Williams, M.L., "The Stresses Around a Fault or Crack in Dissimilar Media", *Bulletin of the Seismological Society of America*, 49, (2), 1959, 199-203
2. England, A.H., "A crack between dissimilar media", *Journal of Applied Mechanics*, 32, 1965, 400-402
3. Gdoutos, E.E., Papakaliatakis, G., "Photoelastic Study of a Bimaterial Plate with a Crack along the Interface", *Engineering Fracture Mechanics*, 16, (2), 1982, 177-187
4. Qu, J., Bassani, J.L., "Cracks on Bimaterial and Bicrystal Interfaces", *Journal of the Mechanics and Physics of Solids*, 37, (4), 1989, 417-433
5. Bassani, J.L., Qu, J., "Finite Crack on Bimaterial and Bicrystal Interfaces", *Journal of the Mechanics and Physics of Solids*, 37, (4), 1989, 435-453
6. Cho, S.B., Lee, K.B., Choy, Y.S., Yuuki, R., "Determination of Stress Intensity Factors and Boundary Element Analysis for Interface Cracks in Dissimilar Anisotropic Materials", *Engineering Fracture Mechanics*, 43, (4), 1992, 603-614
7. Deng, X., "Complete complex series expansions of near-tip fields for steadily growing interface cracks in dissimilar isotropic materials", *Engineering Fracture Mechanics*, 42, (2), 1992, 237-242
8. Deng, X., "General Crack-Tip Fields for Stationary and Steadily Growing Interface Cracks in Anisotropic Bimaterials", *Journal of Applied Mechanics*, 60, 1993, 183-189
9. Sanford, R.J., Dally, J.W., "A general method for determining mixed-mode stress intensity factors from isochromatic fringe patterns", *Engineering Fracture Mechanics*, 11, 1979, 621-633
10. Dally, J.W., Sanford, R.J., "Classification of stress intensity factors from isochromatic fringe patterns", *Experimental Mechanics*, 18, (12), 1978, 441-448
11. Dally, J.W., "Dynamic photoelastic studies of fracture", *Experimental Mechanics*, 19, (10), 1979, 349-361
12. Prabhakaran, R., "Photo-orthotropic elasticity: a new technique for stress analysis of composites", *Optical Engineering*, 21, (4), 1982, 679-688
13. Gdoutos, E.E., Papakaliatakis, G., "Photoelastic Study of a Bimaterial Plate with a Crack along the Interface", *Engineering Fracture Mechanics*, 16, (2), 1982, 177-187
14. Lu, H., Chiang, F.P., "Photoelastic Determination of Stress Intensity Factor of an Interfacia l Crack in a Bi-material", *Journal of Applied Mechanics*, 60, 1993, 93-100
15. Lee, K.H., "Stress and Displacement Fields for Propagating the Crack Along the Interface of Dissimilar Orthotropic Materials Under Dynamic Mode I and II Load", *Journal of Applied Mechanics*, 67, 2000, 223-228
16. Hyer, M.W., Liu, D.H., NASA Rep. 3773, 1984
17. Haixiao, L., Chuhan, Z., Zheng, L., Xtreng, Z., "Photoelastic Studies for Composite Dynamics", *Journal of Materials Science*, 34, 3479-3487, 1999
18. Singh, R.P., Shukla, A., "Characterization of isochromatic fringe patterns for a dynamically propagating interface crack", *International Journal of Fracture*, 76, 1996, 293-310
19. Singh, R.P, Lambros, J., Shukla, A., Rosakis, A.J., "Investigation of the mechanics of intersonic crack propagation along a bimaterial interface using coherent gradient sensing and photoelasticity", *Proceedings of the Royal Society of London, A*, 453, 1997, 2649-2667
20. Akisanya, A.R., Fleck, N.A., "Interfacial Cracking from the Free-Edge of a Long Bi-Material Strip", *International Journal of Solids and Structures*, 34, (13), 1997, 1645-1665
21. Soh, A-K., "An improved photo-elastic technique for determining the mixed-mode stress-intensity factors for interfacial cracks in a bi-material", *Composites Science and Technology*, 59, 1999, 1033-1039
22. Kavaturu, M., Shukla, A., "Dynamic Fracture Criteria for Crack Growth Along Bimaterial Interfaces", *Journal of Applied Mechanics*, 65, 1998, 293-299
23. Shukla, A., "High-speed fracture studies on bimaterial interfaces using photoelasticity – a review", *Journal of Strain Analysis for Engineering Design*, 36, (2), 2001, 119-142
24. Coker, D., Rosakis, A.J., Needleman, A., "Dynamic crack growth along a polymer composite – Homalite interface", *Journal of the Mechanics and Physics of Solids*, 51, 425-460, 2003.
25. Khanna, S.K., Ellingsen, M.D., Winter, R.M., " "Investigation of Fracture in Transparent Glass Fiber Reinforced Polymer Composites Using Photoelasticity", *ASME Journal of Engineering Materials and Technology,* 126, (1), 2004, 1-7
26. Chen, T.Y., "Digital Fringe Multiplication in Three-Dimensional Photoelasticity", *Journal of Strain Analysis*, 30, (1), 1-7, 1995

Proceedings of IMECE04
2004 ASME International Mechanical Engineering Congress and Exposition
November 13-20, 2004, Anaheim, California USA

IMECE2004-61777

FE CALCULATIONS OF *J*-INTEGRALS ON CRACK LENGTH IN A CONSTRAINED ELASTOMERIC DISC WITH CRACK SURFACE PRESSURES AND ISOTHERMAL LOADS

H. K. Ching
Materials Technology Center
Southern Illinois University
Carbonale, IL 62901, U. S. A

C. T. Liu
AFRL/PRSM
10 E. Saturn Blvd.
Edwards AFB, CA 93524, U. S. A.

S. C. Yen
Materials Technology Center
Southern Illinois University
Carbonale, IL 62901, U. S. A.

ABSTRACT

In this study, we performed linear and nonlinear FE (Finite Element) analyses to compute *J*-integrals for a centrally perforated star-shaped disc, which is made of an elastomeric material, under crack surface pressures and isothermal loads. Deformations of the disc were constrained by a circular steel ring enclosing the disc. Different crack sizes were assumed to exist in the front of the star-shaped notches. For linear the analysis, material compressibility was modeled with Poisson's varying form 0.48 to 0.4999. In addition, with the presence of the crack surface pressure, the *J*-integral needs to be modified by including an additional line integral. Numerical studies showed that the value of the *J*-integral increases with the increase of the crack length, reaches a maximum value at 1*in* of crack length, and then decreases gradually. Both linear and nonlinear analyses agree qualitatively but differ quantitatively. It was also found that values of the *J*-integral strongly depend upon the material compressibility.

INTRODUCTION

Defects such as flaws and cracks may form in elastomeric materials due to the manufacturing, handling or ageing. To ensure the integrity and reliability for such structural components, fracture toughness should be ascertained so that the onset of the crack growth can be determined based on the fracture resistance of the material. The energy release rate is a measure of the fracture toughness, and commonly used as a criterion to determine the maximum operating loads for a given pre-existing defect. Most elastomeric materials such as rubbers and solid propellants exhibit mechanical behavior that remains nonlinearly elastic at large strains and have very little compressibility, and hence these materials are often referred to as fully or nearly incompressible. When these elastomers are loaded in a highly confined state, even a small change in the compressibility can result in dramatic difference of stress distributions. For example, Schapery [1] conducted an experiment for a circular polymeric disk under the hydrostatic tension, and showed that a small change in Poisson's ratio can alter the stress distribution significantly.

Many investigations have been conducted to determine the relationship between the crack-tip stress and strain fields and the energy release rate for rubber-like materials. Thomas [2] was the first to study this relationship experimentally. He found that the average strain energy density in a sheet of rubber was uniquely related to the energy release rate regardless of the specimen type. Thomas's conclusion had been validated later by Andrews [3] and Knauss [4]. Andrews used a microscopic, photoelastic technique to quantify the strain fields around the crack tip whereas Knauss used a printed-grid technique. Morman et al. [5] also gave an analytical solution relating the energy release rate to the crack tip radius.

Rice's development of the *J*-integral [6] gave a mathematical argument to characterize the local stress-strain field around a crack front. The *J*-integral was found to be equivalent to the energy release rate and independent of the path contours. Based on the *J*-integral, several path-independent integrals have been proposed for more general elastic-plastic problems by including non-proportional loading and unloading, thermal strains, and material inhomogeneity. A review about the limitations and salient features of these integrals can be found in Kim and Orange's work [7]. Note that Rice's original form of the *J*-integral is valid even for the nonlinear elastic materials. The problem of a crack in an infinite, thin, and incompressible sheet subjected to a biaxial tension at infinity was studied, within the framework of nonlinear elasticity for a Neo-Hookean material, by Wong and Shield [8], and Chang [9] generalized the *J*-integral for nonlinear elastic materials with finite strains.

The presence of the crack surface tractions, which may be due to ignition pressures or pressurized fluids, is of practical interests in engineering applications. In this regard, the *J*-

integral needs to be modified by including an additional line integral so that the path-independence of the *J*-integral can be preserved. Chang and Becker [10] scrutinized the effect of non-conservative surface tractions applied to the crack faces on the energy release rate for a 2D rubber-like material. They pointed out that the usual expression for the energy release rate needed to be modified, and also demonstrated that the energy release rate depends upon the constitutive relation used to model the material response.

In this study, a finite element (FE) analysis was conducted to compute the *J*-integral for a circular elastomeric disc with a pre-existing crack. The disc had a star-shaped hole with six symmetrical notches emanating from the disc's center and was enclosed by a thin steel ring. We assumed that the disk was modeled either as a linearly elastic Hookean material or a nonlinear elastic Ogden material, and the steel ring was modeled as a Hookean material. Material constants obtained from a uniaxial tensile test data were used to describe the Ogden strain energy potential. Three loading conditions, internal pressure, constant thermal load, and combined internal pressure and constant thermal load, were considered. To account for the material compressibility, Poisson's ratio of the Hookean material was assumed to vary from 0.48 to 0.4999.

FORMULATION OF THE PROBLEM

Figure.1 shows a cross-section of the specimen in which an elastomeric disc is enclosed by a thin steel ring. A plane strain state of deformation was assumed to prevail in the body. The star-shaped hole at the center of the specimen had six symmetrically located notches. In order to obtain the *J*-integral as a function of the crack length, a starter crack with 11 different crack sizes, a = 0.05, 0.1, 0.2, 0.4, 0.6, 0.8, 1.0, 2.0, 5.0, 10.0, 15 *in*, was assumed to exist directly ahead of the notch tip. The elastomeric disc and the steel casing were modeled as isotropic and homogeneous materials, and the whole specimen under three types of loading, namely, the pressure load in which the pressure with p = 1000 *psi* is applied on the inner surface of the hole, the isothermal load in which the temperature with T = -85°F is uniformly lowered on the entire body, and the combined pressure and isothermal load, were considered. Deformations of the steel ring are assumed to be infinitesimal and its isotropic material modeled by Hooke's law with Young's modulus = 29 M*psi*, Poisson's ratio = 0.3, and the coefficient of thermal expansion = $6.5\times10^{-6}/\,^{\circ}$F

We performed the FE analysis using the commercial ABAQUS computer code [11] to calculate the *J*-integral for the problem studied. Due to the symmetry of the specimen geometry and loading conditions, only a 30° sector of the specimen was investigated. Figure 2 exhibits the finite element mesh of the sector for a crack size with 1*in* long, and 4824 8-node quadrilateral elements are used with a dense mesh near the crack tip. Because the elastomeritc material was considered to be fully or nearly incompressible, hybrid elements were adopted in the analysis. Points on the bounding surfaces $\theta = 0^{\circ}$ and $\theta = 30^{\circ}$ were constrained to move radially, and tangential tractions on these surfaces were set equal to zero. The outer surface of the steel ring was taken to be traction free. The elastomeric disc was assumed to be perfectly bonded to the steel ring so that displacements and surface tractions were continuous across their common interface.

The constitutive behavior of the elastomeric disc was determined by either the Hooke's law for the linear analysis or the Ogden strain energy potential for the nonlinear analysis. The stress-strain curve of the elastomer was obtained by conducting the uniaxial tension test at a constant displacement rate of 200 *in/min*. For the Ogden material with fully incompressibility, the strain energy potential is

$$U = \sum_{i=1}^{N} \frac{2\mu_i}{\alpha_i^2}(\bar{\lambda}_1^{\alpha_i} + \bar{\lambda}_2^{\alpha_i} + \bar{\lambda}_3^{\alpha_i} - 3) \qquad (1)$$

where $\bar{\lambda}_i$ are the deviatoric principal stretches $\bar{\lambda}_i = J^{-\frac{1}{3}}\lambda_i$; λ_i are the principal stretches; and μ_i, α_i are temperature-dependent material parameters. J is the total volume ratio.

The shear modulus μ_0 at zero strain for the Ogden material can be given by

$$\mu_0 = \sum_{i=1}^{N} \mu_i \qquad (2)$$

The elastic volume ratio, J^{el} is related to J by

$$J^{el} = J/(1+\alpha T)^3 \qquad (3)$$

We set N equal to 2 in the Ogden strain energy and ABAQUS computer code determined the material constants μ_i, α_i from the uniaxial test data through a least-squares-fit procedure. Values of μ_1 and μ_2 determined by ABAQUS are

$$\mu_1 = -160.4 psi,\ \mu_2 = 1643 psi \qquad (4)$$

Computed values of the axial component of the first Piola-Kirchhoff stress vs. the nominal axial strain are compared with the experimental data in Fig. 3. It is clear that the two sets of data are very close to each other implying that values given in Eq. (4) of μ_1 and μ_2 are very accurate.

For the linear analysis, the elastomer was assumed to obey the Hooke's law with Poisson's ratio ν varying from 0.48 to 0.4999. The stress-strain curve for the Hookean material is also depicted in Fig. 3. The Young's modulus is set equal to the slope of the stress-strain at zero strain and can be computed by

$$E = 2(1+\nu)\mu_0 \qquad (5)$$

where μ_0 is obtained from Eq. (2).

For both linear and nonlinear analyses, the coefficient of thermal expansion for the disk material equaled $5.6\times10^{-6}/\,^{\circ}$F.

When the inner surface of the hole and the surface of the starter crack, aligned with the $x_1 - \text{axis}$, are loaded by a uniform pressure p, the J-integral for an elastic material is given by

$$J = \int_{\Gamma}\left(Wn_1 - T_{ij}n_j \frac{\partial u_i}{\partial x_1} ds \right) + \int_{\Gamma_c} p \frac{\partial u_2}{\partial x_1} dx_1 \quad (6)$$

where W is the strain energy density, **u** is the displacement of a point, Γ is a closed curve enclosing the crack tip, **n** is the outward unit normal to Γ, and T_{ij} is the stress tensor, and Γ_c are the two crack faces. Note that the second term on the right-hand side of (6) represents the work done by the pressure on the crack surfaces. For the thermal load, there is no surface traction on the crack surface and this term makes null contribution to the value of J. For a homogeneous hyperelastic material, T_{ij} equals the first Piola-Kirchhoff stress tensors.

The J-integral can be related to the stress intensity factor K by the Irwin's relation [12]. As the crack is subjected to the pressure load or isothermal load, only mode I deformations exist around the crack. Thus, the J-integral can be written as

$$J = \frac{1-\nu^2}{E} K_I^2 \quad (7)$$

RESULTS AND DISCUSSIONS

Figures 4(a) and 4(b) show the dependence of the J-integral on the crack length for the linear and nonlinear analyses under the pressure and thermal loads, respectively. Poisson's ratio in the linear analysis is set equal to 0.4999 in order to compare results with those obtained from the nonlinear analysis. The two analyses agree each other qualitatively but differ quantitatively. In both loading conditions, the J-integral initially increases with an increase in crack length, reaches a maximum value at a crack length of about 1.0*in*, and then gradually decreases with an increase in crack length. The linear analysis gives larger values of the J-integral than those of the nonlinear analysis for each of the crack length.

The differences of the J-integral in two analyses can be more evident by plotting the maximum principal stress around the crack tip with the crack length of 1*in* for two loads. The contours of the in-plane maximum principal stress are exhibited in a small region of 1in^2 where the crack tip is intensively deformed. As for either the pressure or the isothermal loads, the two sets of the principal stresses look alike but differ a little bit. It is shown that the principal stress induced at a point near the crack tip is compressive for the pressure load while it is tensile for the isothermal load. Note that a small displacement analysis is assumed in the Hookean material and a large deformation analysis is made in the Ogden model, and thus the differences on the value of the J-integral are not only due to the material models considered but also to the geometrical nonlinearity of the specimen.

To investigate the effect of material compressibility on the computed values of the J-integral, four different Poisson's ratios, ν = 0.48, 0.49, 0.499 and 0.4999, are assumed for the material of the elastomeric disc in the linear analysis. Variations of the J-integral with the crack length are depicted in Figs. 6(a), (b) and (c) for the pressure, isothermal and combined loads, respectively. Whereas the results are qualitatively similar to each other for different values of Poisson's ratio, the values of the J-integral vary significantly with the Poisson's ratio under the pressure load. The value of the J-integral for ν = 0.48 reaches as high as 143.3 *lb*/*in*, which is approximately 40 times that with J = 3.557 *lb*/*in* for ν = 0.4999. It is clear that a slight difference in material compressibility can drastically change the value of the J-integral for the case of the pressure load, and henceforth for the combined load (Fig. 6(c)). When the disc is subjected to the isothermal load, the values of the J-integral do not vary much as those with the pressure load.

For the crack length of 1*in*, we further delineate the variation of J-integral with the Poisson's ratio in Fig. 7. The value of the J-integral decreases sharply with the increase of Poisson's ratio from 0.48 to 0.4999 for the pressure load, while it increases gradually with the increase in Poisson's ratio. Recalling that in Eq. (7) the value of J-integral is proportional to K_I, thus one can verify that value of J-integral is proportional to p^2 under the pressure load or T^2 under the isothermal load. As such, even for the linear analysis, values of the J-integral for the combined load cannot be obtained by adding the energy release rates for the corresponding pressure and isothermal loads. The results of the analyses show that, under pressure load, the increase in ν will decrease the compressibility of the material which, in turn, will decrease the crack opening deformations between the two crack faces (Figure 8(a)). For the thermal load, the deformed shape of the crack faces, plotted in Fig 8(b) for different values of Poisson's ratios, does not differ too much. When the Poisson's ratio is equal to 0.4999, the deformed shapes between the Hookean and Ogden materials are almost indistinguishable. The relationship between the J-integral and the crack opening displacement δ can be ascertained in Fig. 9 where 2δ is defined to be the vertical distance between the opposite ending points on the crack faces. It is found that the value of J-integral is also proportional to δ^2. For the crack length of 1*in*, the J-integral can be determined by

$$J = 2.6268 \times 10^3 \delta^2 \quad (8)$$

Therefore, for the present configuration of the elastomeric disk, the value of the J-integral can be obtained by measuring the crack opening displacement when the disc is loaded.

In classical linear elastic fracture mechanics, the near-tip stress fields can be characterized by a single parameter such as the J-integral or the stress intensity factor. However, previous studies have shown that the T-stress is also an important parameter in describing the state of stresses near the crack tip. For example, Larsson and Carlsson [13] showed that the sign and magnitude of the T-stress have a significant effect on the size and shape of a plastic zone around the crack tip. Kirk et al. [14] and Sorem et al. [15] showed that the fracture toughness of a given material can be considerably dependent on the T-stress. It has also been shown by Cotterel and Rice [16] that the T-stress plays an important role in the crack path stability. Figure 10 depicts the variation of T-stress with the crack length for the pressure and isothermal loads when the Poisson's ratio is equal to 0.4999. For both loading cases, the magnitude of the T-stress decreases with the increase of the crack length, and the T-stress is in compression (i.e. negative values of T-stress). Betegon and Hancock [17] have indicated that when the T-stress corresponds to the negative value, crack-tip deformations are dominated by the combination of the J-integral and T-stress. Furthermore, the fracture trajectory is path stable for a cracked specimen with negative T-stress. The dependence of the T-stress on the Poisson's ratio is evinced in Fig. 11. Similar to the J-integral, the value of the T-stress strongly depends on the Poisson's ratio for the pressure load while it does not for the isothermal load. It can be easily verified that T-stress for the combined load can be obtained by adding the values of T-stress from the corresponding pressure and isothermal loads.

CONCLUSIONS

Linear and nonlinear finite element analyses were conducted to obtain the variations of the J-integral with the crack length for an elastomeric disk enclosed by a steel ring. For both analyses, the value of the J-integral increases with an increase in the starter crack length, reaches a maximum value at $1in$ of crack length, and then decreases gradually. It was shown that, for different loading conditions and each one of the crack length considered, the analysis of the linear problem gives higher values of the J-integral than those for the nonlinear problem. Material compressibility of the disc is also considered by varying the Poisson's ratio from 0.48 to 0.4999 for the Hookean material. For the pressure load, the value of the J-integral strongly depends upon the values of the Poisson's ratio. This is due to the reason that the work done by the crack surface pressure increases noticeably as the value of the Poisson's ratio is decreased, resulting in a significant increase in the J-integral. For the thermal load, the value of J-integral does not differ much with the variation of the Poisson's ratio. It is interesting and important to point out that, even for the linear analysis, the value of J-integral for the combined load cannot be obtained by adding that of the J-integral obtained from the pressure and the thermal loads separately. Dependence of the T-stress on the crack length and material compressibility was also examined. The magnitude of the T-stress decreases with the increase in crack length, and the T-stress is found to be in compression.

The problem studied herein can also be viewed as the solid propellant grains cast into a metallic casing under the ignition pressure or thermal shock loads. It is concluded that a proper value assigned to the Poisson's ratio is important in evaluating the fracture toughness of the solid rocket motor under the pressure load. An inaccurate value of the Poisson's ratio may result in a wrong evaluation on the structural integrity of a missile system.

ACKOWLEDGEMENTS

The first author would like to acknowledge the support from the Air Force Research Laboratory and Materials Technology Center at Southern Illinois University in Carbondale.

REFERENCES

1. Lindsey, G. M., 1967, *J. Appl Phys*. 38(12), pp. 4843-4852.
2. Thomas, A. G., 1955, "Rupture of rubber. II. The strain concentration at an inclusion", *J. Polymer Science* 18, 177-188.
3. Andrew, E. H., 1963, "Rupture propagation in hysteresial materials: stress at a notch." *J. Mechanics and Physics of Solids* 11, pp. 231-242.
4. Knauss, W. G., 1968, "Stress near a crack in a rubber sheet", *Experimental Mechanics* 8, pp. 177-181.
5. Morman, K. N., Mazich, K. A., Oblinger, F. G., Zhang, F., and Killgoar, P. C., 1992, "Critical tearing energy in a circumferentially-cracked circular cylinder of rubber under finite deformation", *Int. J. Fracture* 53, pp. 129-157.
6. Rice J. R., 1968, "A path-independent integral and the approximate analysis of strain concentration by notches and cracks", *J. Applied Mechanics* 35, pp. 379-386.
7. Kim, K. S. and Orange, T. W., 1988, "A review of Path-Independent integrals in elastic-plastic fracture mechanics", *American Society for Testing and Materials*, Philadelphia, pp. 713-729.
8. Wong, F. S., and Shield, R. T., 1969, "Large plane deformations of thin elastic sheets of neo-Hookean material", *Z. Angew. Math. Phys*. 20, pp. 176-199.
9. Chang, S. J., 1972, "Path-independent integral for rupture of perfectly elastic materials", *J. Applied Math. and Phys.* 23, pp.149-152.
10. Chang, J. H., and Becker, E. B., 1992, "Finite element calculation of energy release rate for 2-D rubbery material problems with non-conservative crack surface tractions", *Int. J. Num. Math. Eng*. 33, pp. 907-927.
11. ABAQUS/Standard user's manual 6.2, 2001, Hibbitt, Karlsson & Sorensen, Inc., Providence, RI.
12. Irwin, G. R., 1948, "Fracture dynamics. In fracturing of metals", *American Society of Metals*, pp. 147-166.
13. Larsson, S. G. and Carlsson, A. J., 1973, "Influence of non-singular stress terms and specimen geometry on small scale yielding at crack tips in elastic-plastic materials", *J. Mechanics and Physics of Solids*, 21, pp. 263-277.
14. Kirk, M. T., Koppenhoefer, K. C. and Shih, C. F., 1988, "Effect of constraint on specimen dimensions needed to obtain structurally relevant toughness measures", *American Society for Testing and Materials*, Philadelphia, pp. 79-103.
15. Sorem, W. A., Dodds, R. H. and Rolfe, S. T., 1991, "Effects of crack depth on elastic plastic fracture toughness", *Int. J. Fracture* 47, pp. 105-126.
16. Cotterell, B. and Rice, J. R., 1980, "Slightly curved or kinked crack", *Int. J. Fracture* 16, pp. 155-169.
17. Betegon, C. and Hancock, J. W., 1991, "Two-parameter characterization of elastic-plastic crack-tip fields", *J. Applied Mechanics*, 58, pp. 104-110.

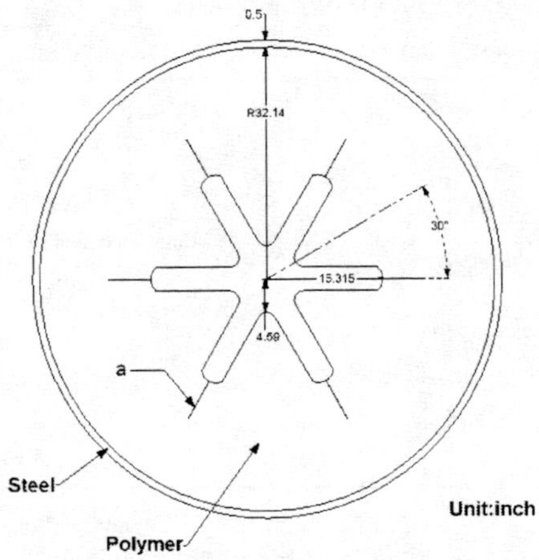

FIGURE 1. Specimen geometry

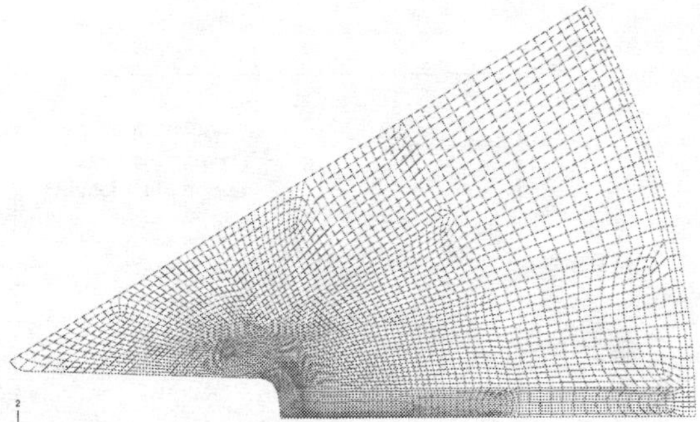

FIGURE 2. For 1*in* long crack, the finite element mesh with 4824 elements for the analysis of the problem

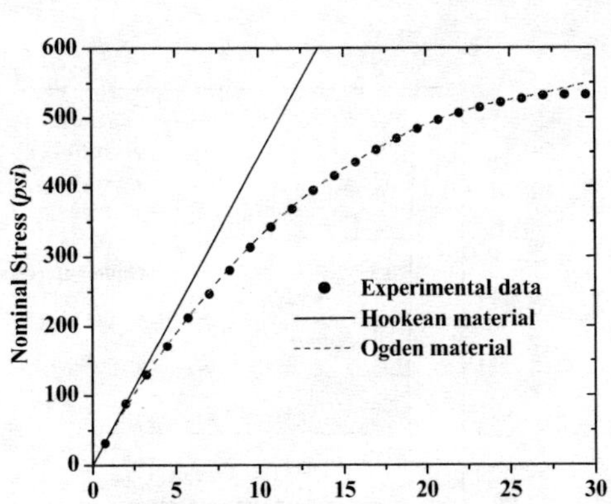

FIGURE 3. A comparison of the computed and the experimental axial nominal stress vs. nominal strain for the disk material

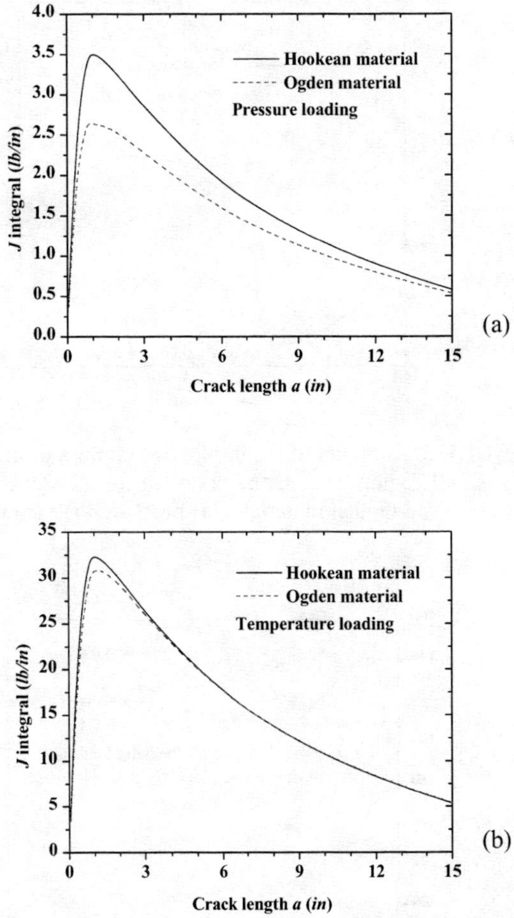

FIGURE 4. Variation of the *J*-integral with the crack length when the disk material is modeled either as a Hookean material or an Ogden material; (a) pressure load (b) isothermal load

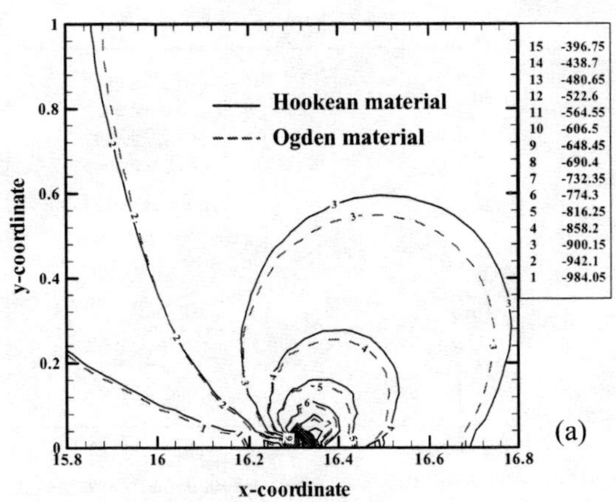

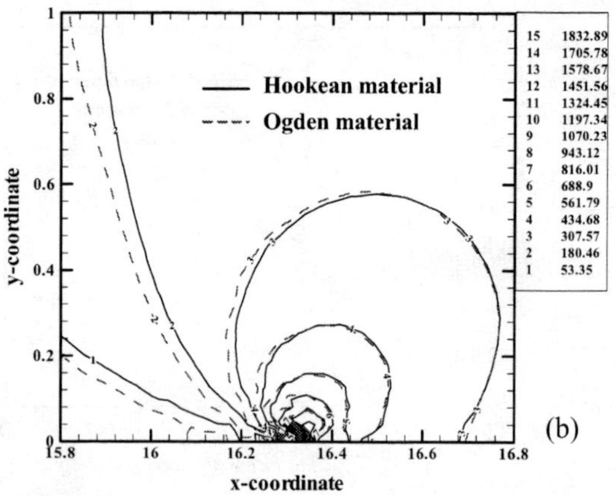

(b)

FIGURE 5. Contours of the in-plane maximum principal stress in a small region around the crack tip of 1*in* long crack for Hookean and Ogden materials; (a) pressure load (b) isothermal load

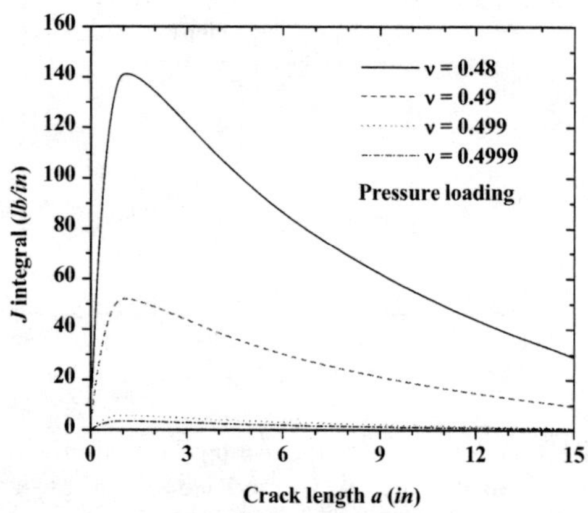

(a)

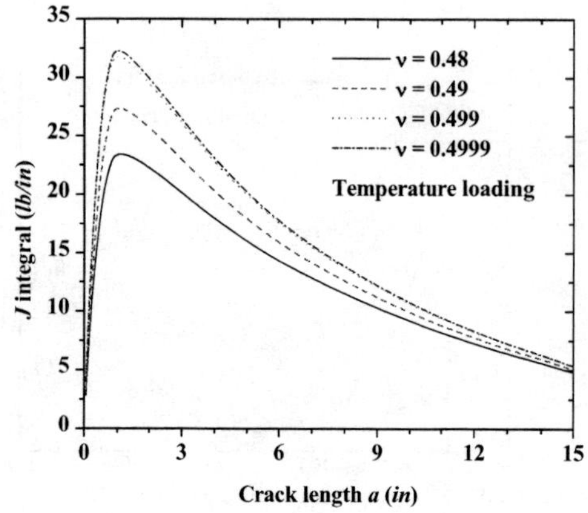

(b)

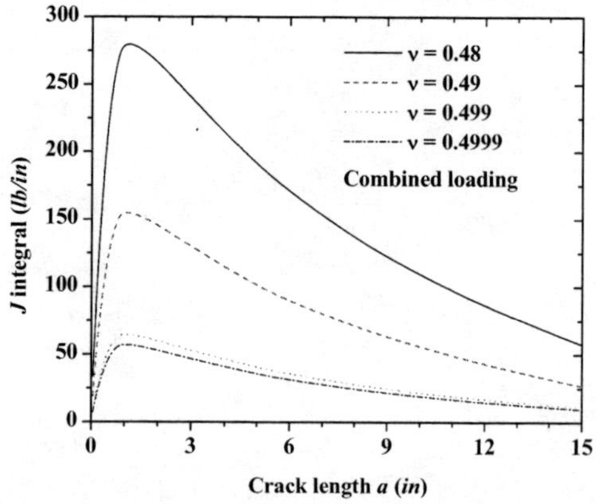

(c)

FIGURE 6. Variation of the *J*-integral with the crack length for different values of Poisson's ratio; (a) pressure load (b) isothermal load (c) combined load

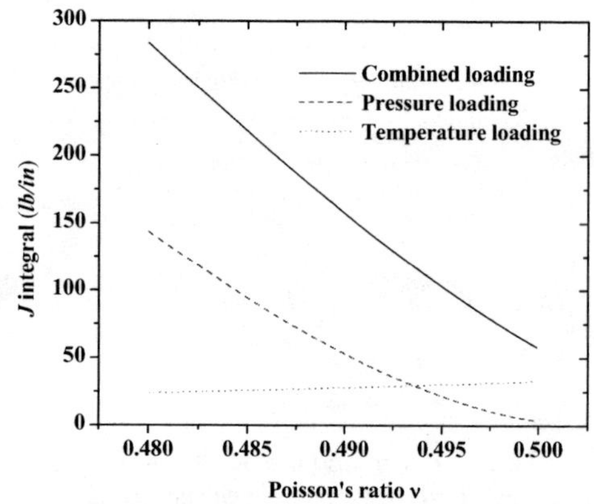

FIGURE 7. The dependence of the *J*-integral upon Poisson's ratio for three loadings

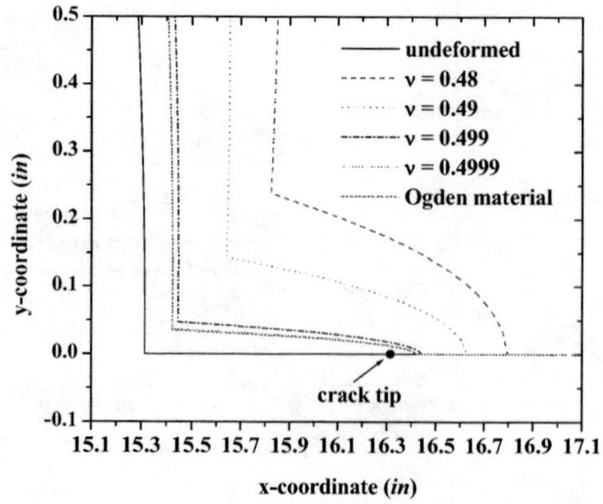

(a)

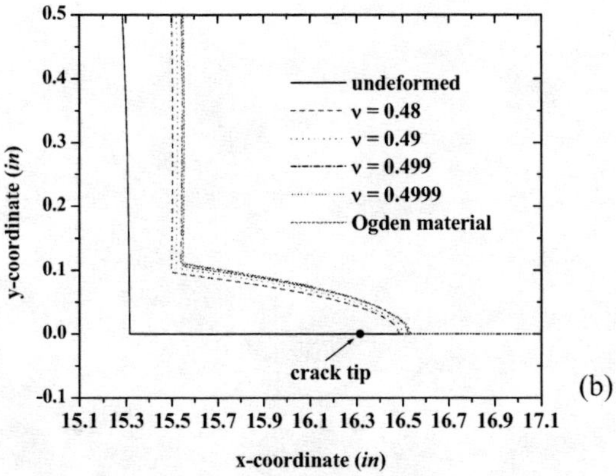

FIGURE 8. Deformed shapes of the crack face for the Ogden material and Hookean material with different values of Poisson's ratio; (a) pressure load (b) isothermal load

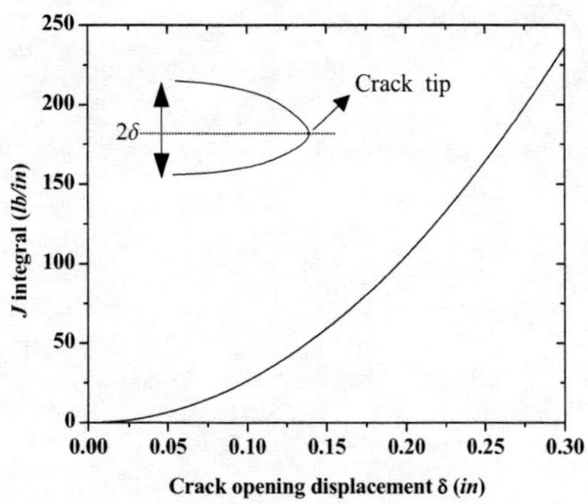

FIGURE 9. For 1*in* long crack, variation of the *J*-integral with the crack opening displacement

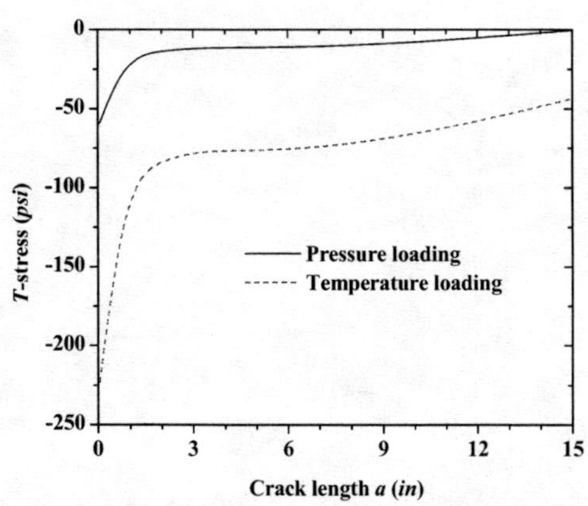

FIGURE 10. Variation of the *T*-stress with the crack length for the pressure and isothermal load

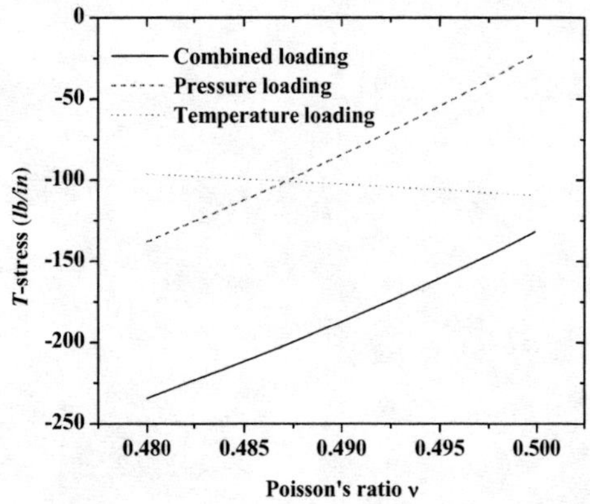

FIGURE 11. The dependence of the *T*-stress upon Poisson's ratio for three loadings

Proceedings of IMECE04
2004 ASME International Mechanical Engineering Congress and Exposition
November 13-20, 2004, Anaheim, California USA

IMECE2004-62146

Multi-axial fatigue damage models of fiber reinforced composites

N. H. Yang, H. Nayeb-Hashemi* and **A. Vaziri+**
Department of Mechanical, Industrial and Manufacturing Engineering, Northeastern University, Boston, MA 02115

ABSTRACT

Fiberglass reinforced composites are extensively used in various structural components. In order to insure their structural integrity, their monotonic and fatigue properties under multiaxial stress fields must be understood. Combined in-phase tension/torsion loading is applied to $[\pm 45^o]_4$ E-glass/epoxy composite tubes under monotonic and fatigue conditions to determine the effects of multiaxial loading on its failure. Various monotonic and fatigue damage criteria are proposed. These models considered failure mode (failure plane), the energy method and the effective stress-strain method. It is observed for the majority of experiments, the failure initiated at the outer lamina layer at 45° to the tube axis. A damage criterion for multiaxial monotonic loading is proposed considering both normal and shear stress contributions on the plane of failure. The experimental data show an excellent agreement with this proposed model for various loading conditions. Other failure models are currently under investigation utilizing the stresses and strains at the composite laminate as well as stress and strain at the outer lamina layer.

Multiaxial fatigue failure models are proposed considering again the plane of failure. Since the plane of the failure is subjected to mean and cyclic stresses (shear and normal) and mean and cyclic strains (shear and normal), the fatigue damage models consider the contributions of these stresses and strains to the fatigue life of the composite tube. In addition to the fatigue damage model based on the plane of failure, a multi-axial fatigue failure model is proposed considering the mean and cyclic energy during fatigue experiments. The experimental data show a good correlation between the proposed damage parameters and fatigue life of specimens with some scatter of the data. Other fatigue failure models are currently under investigation considering the loading frequency and visco-elastic properties of the composite.

INTRODUCTION

Fiber reinforced composites are extensively used in manufacturing of various components such as high-pressure vessels, aerospace structures, transmission shafts in automobiles, support structures, etc. Traditional materials are being replaced by composites due to their high strength to weight ratio, corrosion resistance and cost. Increase use of composite structures has led to the need for models to determine their monotonic and their fatigue characteristics under multiaxial stress fields. There are numerous papers reporting fatigue damage models in isotropic materials under multiaxial fatigue loadings. However, these models are not suitable for evaluating fatigue life of anisotropic fiber reinforced composites.

There is relatively little literature on the damage accumulation and the effect of multi-axial loading on the fatigue life of composite materials. Found [1] proposed a conservative failure criterion for chopped fiber reinforced composites, woven fabric reinforced composite and fabric reinforced composites under bi-axial static and fatigue loading conditions. He suggested a conservative estimate for design purposes where the failure in the tension/tension quadrant is given by a circular arc and by a straight line in the tension/compression quadrant. The proposed failure models are given as,

$$\left(\frac{\sigma_1}{F_{1t}}\right)^2 + \left(\frac{\sigma_2}{F_{2t}}\right)^2 = 1 \quad (1)$$

$$\left(\frac{\sigma_1}{F_{1t}}\right) - \left(\frac{\sigma_2}{F_{2c}}\right) = 1 \quad (2)$$

where σ_1 and σ_2 represent stresses in the principal direction, F_{1t} and F_{2t} are tensile strength in the principal direction and F_{2c} is

* corresponding author; hamid@coe.neu.edu
+ current address: Division of Engineering and Applied Sciences, Harvard University, Cambridge, MA, 02138

compressive strength in the material principal direction. Radon and Wachnicki [2] studied the fatigue crack growth in a chopped strand mat glass fiber reinforced polyester resin under biaxial loading. The fatigue crack growth rate was evaluated through the changes in the specimen compliance. It was shown that the fatigue crack growth in this material could be governed by a Paris law relationship. The micro-mechanisms of the crack growth were through accumulation of various damage modes such as fiber breakage, matrix crazing and layer debonding. Jones et al [3] also investigated the monotonic and fatigue properties of composite materials under biaxial loadings. However, they did not present a model to predict the fatigue life of the specimens.

In contrast to the multiaxial fatigue research reports on composite material, there are a number of reports on the fatigue properties of fiber reinforced composites under uni-axial loading. Kaynak and Mat [4] performed uni-axial fatigue tests on [±55°] glass/epoxy composite tubes. They evaluated the effect of stress level and especially loading frequency on the different stages of failure. It was observed that fatigue damage mechanism consists of three stages. First, craze formation in the matrix material, followed by its densification along filament winding and finally fiber breakage. Krempl and Niu [5] conducted tests on graphite/epoxy [±45°] tubes. Experiments were performed under static conditions and under completely reversed load controlled axial loading at frequencies of 0.01, 0.1 and 1.0 Hz. However, no failure model was proposed. Caprino [6] developed a model to predict the fatigue life of randomly oriented glass fiber reinforced plastics under tension-tension fatigue and applied it to experimental data available in literature. The fatigue life of a specimen is presented as,

$$N = \left[1 + \frac{1}{\alpha(1-R)}\left(\frac{\sigma_o}{\sigma_{\max}} - 1\right)\right]^{1/\beta} \qquad (3)$$

where α and β are constants, R is the stress ratio ($\sigma_{min}/\sigma_{max}$), σ_o is the monotonic strength of the virgin material and $\sigma_{\max}$ is the maximum stress during cyclic loading.

There is substantially more literature on the prediction of multiaxial fatigue failure of metals and metal alloys. Under cyclic loading, metals and metal alloys differ from composites in that metals tend to fail with the growth of a single crack where as composites fail from an accumulation of various damages. Metals also exhibit flow and hardening factors that must be accounted for in multi-axial fatigue models [7, 8]. However, the methods proposed to predict fatigue failure in metals may provide insights to developing damage criteria and fatigue life of composites.

There are three commonly used methods to predict multiaxial fatigue failure in isotropic material: critical plane of failure method, work/energy method and effective stress-strain method. The critical plane method utilizes the normal and shear stresses and strains at the plane of failure as damage parameters in the fatigue failure model. Apparently there are two planes of failure [9], case A, where the critical plane of failure is defined as the one that crack propagates along the surface of the specimen, and case B, where the crack grow away from the surface of the specimen and into the specimen. Based on these two planes, the fatigue damage parameter is defined as,

$$\gamma^* + k\varepsilon^* \qquad (4)$$

as a contour of constant life. Where k is a constant, γ^* is the shear strain amplitude and ε^* is the normal strain amplitude to desired critical plane. Socie et al [10] modified the above equation based on multi-axial fatigue tests on Inconel 718. The fatigue damage model was defined as,

$$\gamma_p^* + k\varepsilon_{np}^* \qquad (5)$$

where γ_p^* and ε_{np}^* are the plastic shear strain amplitude and the plastic strain amplitude normal to the critical plane for growth of the crack into the specimen, respectively. The above models do not consider the effect of the mean stress and strain on the critical plane of failure into the damage model. Socie modified these damage models by incorporating the effect of normal mean stress on the critical plane into the damage model as,

$$\gamma_p^* + k\varepsilon_{np}^* + \frac{\sigma_{no}^*}{E} \qquad (6)$$

where σ_{no}^* is the mean normal stress at the critical plane. Fatemi and Socie [11] proposed a fatigue damage parameter based on the plane of the maximum shear and its normal stress as,

$$\gamma_{\max}\left(1 + \frac{n\sigma_n^{\max}}{\sigma_y}\right) \qquad (7)$$

where $\gamma_{\max}$ is the maximum shear strain at the critical plane, $\sigma_n^{\max}$ is the maximum normal stress at the critical plane , σ_y is the material yield strength and n is a constant.

The energy method reflects the measure of the amount of energy required to cause failure. Ellyin et al [12-15] has studied the multi-axial fatigue damage for a variety of metal alloys. A damage model based on cyclic plastic strain energy was defined as,

$$\Delta W_p = \Delta\bar{\sigma}\Delta\bar{\varepsilon}_p \qquad (8)$$

where $\Delta\bar{\sigma}$ is the equivalent stress range and $\Delta\bar{\varepsilon}_p$ is the equivalent plastic strain range, both based on the von Mises yield criterion. Ellyin and Kujawski [16] proposed alternative damage criteria considering the effects of mean stress on the cyclic plastic strain energy as,

$$\psi = \frac{\Delta W_d h(\sigma_m)}{f(\bar{\rho})} \qquad (9)$$

where ΔW_d is the distortion strain energy range, $f(\bar{\rho})$ is a function of the multi-axial constraint and $h(\sigma_m)$ is a function of the tensile mean stress.

The effective stress-strain method reduces a three dimensional stress state to an effective stress state parameter. Leese and Morrow [17] proposed a fatigue failure model from torsional low cycle fatigue tests on thin-walled steel tubular specimens using the effective shear strain approach. Another model often referred to as SWT [18] model is presented as,

$$\sigma_1^{max}\frac{\Delta\varepsilon_1}{2} = f(N) \tag{10}$$

where the failure plane is defined as the one that has highest $\sigma_1^{max}\frac{\Delta\varepsilon_1}{2}$, where σ_1^{max} is the maximum normal stress and $\Delta\varepsilon_1$ is the cyclic strain range.

Based on the literature survey, there are not extensive studies on the development of a multi-axial fatigue failure model for fiber reinforced composite. In this study, the failure properties of E-glass/epoxy $[\pm 45^o]_4$ tubes under combined tension and torsion are obtained and various damage model criteria for composite laminates under multiaxial loadings are proposed. The goal of this paper is to develop a damage criterion that can be used in a multi-axial fatigue failure prediction. Similar to many other engineering structures, the composite under investigation has anisotropic strength properties which must be accounted for in the damage criteria. The damage criterion must also be applicable for all applied loading conditions. An extension of the three methods previously developed for isotropic material is utilized to derive damage criterion for the composite structure in order to determine which fatigue damage model best fit our experimental data.

EXPERIMENTAL INVESTIGATIONS

The material used in this investigation is an E-glass/epoxy composite provided by Aldila, Inc. The material was provided in a form of unidirectional prepreg with fiber volume friction of 55% and lamina thickness of 0.158 mm. The typical mechanical properties of the unidirectional composite are listed in Table 1. These mechanical properties were obtained using ultrasonic method [19]. The composite material is stored in a freezer as a rolled sheet 609.6 mm wide. The composite is taken out of the freezer where it is allowed to reach room temperature before production of test specimens.

The specimens are hand made by wrapping 8 layers of prepreg around an aluminum mandrill with diameter of 25.4 mm and 304.8 mm long at $\pm 45^o$ to the mandrill axis. Prior to manufacturing composite tubes, the mandrill was cleaned with Xtend CX-500 mold cleaner and then generously coated with Xtend SAM-19 mold release. A six inch wide, thin aluminum sheet is wrapped around the composite layers and then shrink wrap tape provided by Dunstone Company, Inc. is wrapped around the entire specimen before it is cured in an oven at 180° C for 90 minutes. The mandrill is removed once the specimen returns to room temperature. The product of this process is a composite specimen of $[\pm 45^o]_4$ E-glass/epoxy composite tube with an inner diameter of 25.4 mm and an outer diameter of 27.94 mm approximately 304.8 mm long.

Two dumbbell shape inserts are used to make the overall dog bone shape specimen, Fig. 1. In order to insure alignment of the inserts with the specimen axis, a rod is passed through the center of the entire assembly. Furthermore, an aluminum insert with the length of 50.8 mm is inserted between the two dumbbells. The aluminum insert is free and is not carrying any load during experiments. However, it prevents micro-buckling in the composite tube during torsion tests and maintains the tube diameter during tensile tests, thus exerting an internal pressure on the composite tube. The composite tube is glued to dumbbells using ECCOBONDG-909 and is cured at 180° C for 60 minutes. The alignment rod is removed from the assembly prior to experiment.

An Instron tension/torsion machine is used to test the specimen under tension/torsion loading. All monotonic tests are performed under load control of less than 450 N/sec. A data acquisition system is used to continuously monitor load and displacement. All fatigue tests are performed under load control at frequency of 1 Hz. The specimen degradation during test is identified through the changes in the hysteresis curve.

Three different type of multi-axial fatigue testing are performed. The first multi-axial fatigue experiment is performed at a mean tensile stress and fully reversible torsion. The mean tensile stress for all experiments is 119 MPa. Different loading conditions are obtained by adjusting the cyclic tensile and torsion load amplitudes. The second multi-axial fatigue experiment is performed at a constant torsion and cyclically applied tension at a mean load. The last multi-axial fatigue experiment is performed with a constant tensile load with full reversible cyclic torsion.

Table 1 Mechanical properties of the E-glass/epoxy composite

E_1 (GPa)	E_2 (GPa)	G_{12} (GPa)	ν_{12}	Thickness (mm)
53.0	17.65	8.82	0.25	0.158

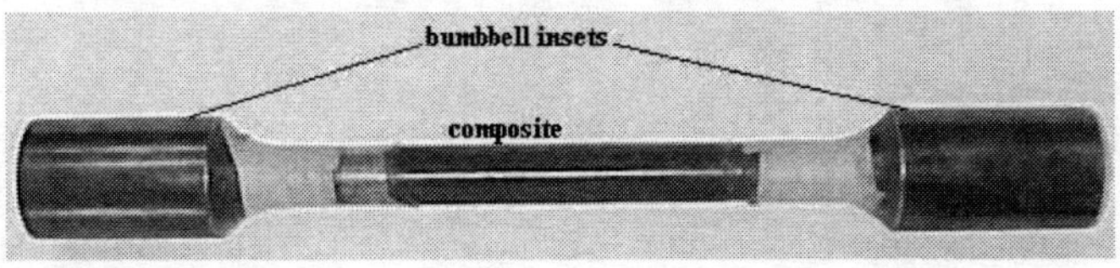

Fig. 1 Composite/dumbbell assembly test specimen

RESULTS AND DISCUSSION

Fig. 2 shows the failure locus of the composite tube under tension/torsion loadings. The results suggest that a quadratic failure criterion could be defined for the composite specimens. However, a closer examination of failed specimens indicates that the failure of all specimens initiated at the outer layer lamina in the 45° plane to the specimen axis, Fig. 3. It is hypothesized that the combination of shear stress and normal stress on this plane may be responsible for the failure load. Based on this assumption a failure model is defined as,

$$\sigma_n + \alpha\tau_n = C \tag{11}$$

where σ_n and τ_n are normal and shear stress at the outer layer surface at the 45° plane and are obtained using the classical lamination theory. Using the classical lamination theory, the strains at the mid surface plane of the composite tube are calculated from the force and torque applied to the tube as,

$$N_y = \frac{F}{2\pi R} \quad (12)$$

$$N_{xy} = \frac{T}{2\pi R^2} \quad (13)$$

where N_y and N_{xy} are the in-plane forces per unit length and R is the radius of the mid surface of the laminate. The strains in the mid surface of the composite are found from the composite lamination theory as,

$$\begin{bmatrix} N_x \\ N_y \\ N_{xy} \end{bmatrix} = \begin{bmatrix} A_{11} & A_{12} & 0 \\ A_{21} & A_{22} & 0 \\ 0 & 0 & A_{66} \end{bmatrix} \begin{bmatrix} \varepsilon_x^o \\ \varepsilon_y^o \\ \gamma_{xy}^o \end{bmatrix} \quad (14)$$

$$A_{ij} = \sum_{k=1}^{N} (\overline{Q}_{ij})_k \, (z_k - z_{k-1}) \quad (15)$$

The $[A]$ matrix represents the stiffness matrix of the composite laminate and $(\overline{Q}_{ij})$ is the component of the stiffness matrix of the individual lamina, which is obtained from the elastic properties of the composite in the material principal direction, and z_k is the location of the lamina in the laminated composite. A MATLAB program is written to evaluate the $\overline{Q}_{ij}$ components of the stiffness matrix of the individual lamina and A_{ij} stiffness of the composite. The normal and shear stress in the 45° plane at the outer layer is subsequently obtained from,

$$\begin{bmatrix} \sigma_x \\ \sigma_y \\ \tau_{xy} \end{bmatrix}_k = [\overline{Q}]_k \begin{bmatrix} \varepsilon_x^o \\ \varepsilon_y^o \\ \gamma_{xy}^o \end{bmatrix} \quad (16)$$

The strain ε_x^o (hoop strain) is assumed to be zero, since the aluminum insert prevents radial displacement. The stress field in the plane of failure is then obtained from the stress transformation as,

$$\sigma_n = \frac{\sigma_x + \sigma_y}{2} + \frac{\sigma_x - \sigma_y}{2}\cos(2\phi) + \tau_{xy}\sin(2\phi) \quad (17)$$

$$\tau_n = -\frac{\sigma_x - \sigma_y}{2}\sin(2\phi) + \tau_{xy}\cos(2\phi) \quad (18)$$

Fig. 4 shows a good agreement between the experimental data and the proposed failure criteria. Other monotonic failure criteria are currently under investigation. One such investigation is based on a proposed time and temperature dependent multi-axial failure model by Richardson et al [20]. This model considers the first stress invariant and the second deviatoric stress invariant as failure criteria.

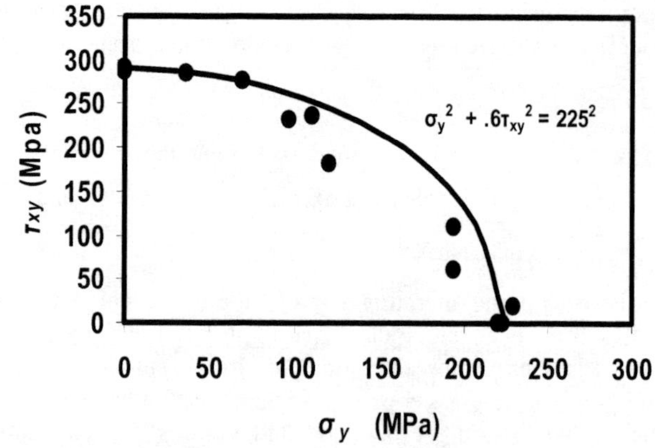

Fig. 2 Monotonic combined loading and failure criteria

Fig. 3 Typical failure mode of a composite specimen in a monotonic multi-axial test

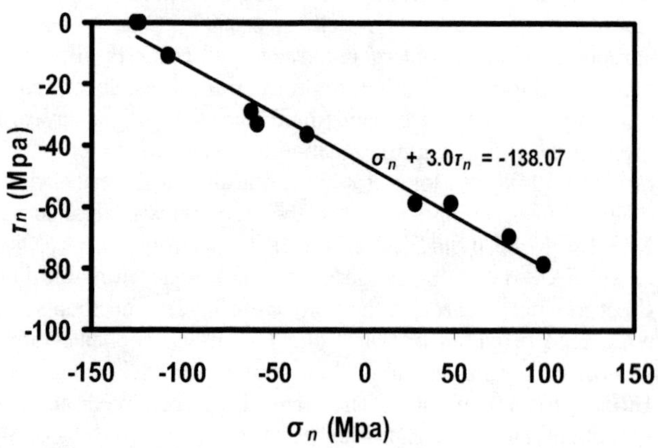

Fig. 4 Failure locus based on the normal and shear stresses on the plane of the failure.

Fatigue Failure Models

Fatigue failure of a majority of the specimens initiated again on the outer surface of the composite laminate at 45° to the specimen axis (Fig 5). Since this plane is subjected to both cyclic and mean normal and shear stresses, a fatigue damage model incorporating both cyclic stress amplitudes and mean stresses is proposed as,

$$D = \frac{(\sigma_n)_a + \alpha(\tau_n)_a + (\sigma_n)_m + \alpha(\tau_n)_m}{C} \qquad (19)$$

where $(\sigma_n)_a$ is the amplitude of the normal stress, $(\tau_n)_a$ is the amplitude of the shear stress, $(\sigma_n)_m$ is the normal mean stress, $(\tau_n)_m$ is the shear mean stress amplitude, α is a material constant and C is the monotonic failure strength defined in eq. (11).

Using again the classical lamination theory, the stress field at the outer layer is evaluated and related to the fatigue life of the specimen using damage model proposed in the eq. (19). The results show the data are somewhat scattered using this damage model, Fig. 6. Since the contribution of mean normal stress is more detrimental to the fatigue life of composite than the mean shear stress, a different weighting factor is considered for mean normal stress than the mean shear stress. This can be justified considering that the mean normal stress reduces crack surface interaction. Considering these arguments, the fatigue damage parameter is defined as,

$$D = \frac{(\sigma_n)_a + \alpha(\tau_n)_a + (\sigma_n)_m + \beta(\tau_n)_m}{C} \qquad (20)$$

where α and β are material constants. Fig. 7 shows that the above damage model predicts fatigue life of the composite tube under various loading conditions reasonably well.

To verify the above damage model, different loading conditions were investigated. The different loading conditions include cyclic tension at a mean tensile load under a constant torsion force and fully reversible cyclic torsion under a constant tensile load. Figure 8 shows that the data for multi-axial fatigue life base on equation (20) are scattered for various loading conditions. Therefore, a modified fatigue damage parameter was defined as,

$$D = \frac{A(\sigma_n + \alpha\tau_n)_a + B(\sigma_n + \beta\tau_n)_m}{C} \qquad (21)$$

where A and B are material constants. Fig. 9 shows that the above damage model is able to predict fatigue life reasonably well for all types of loading conditions.

Furthermore, a damage model was proposed based on the cyclic and mean energy. The energy damage model is similar to the proposed failure model by Ellyin in eq. (8) and is presented as,

$$D = \frac{1}{2}\alpha(\sigma_x\varepsilon_x + \sigma_y\varepsilon_y + \tau_{xy}\gamma_{xy})_{mean} + \frac{1}{2}\beta(\sigma_x\varepsilon_x + \sigma_y\varepsilon_y + \tau_{xy}\gamma_{xy})_{amplitude} \qquad (22)$$

where σ_x, σ_y, τ_{xy} are the average stress in the laminate, ε_x, ε_y and γ_{xy} are the mid-plane strains and α and β are material constants. Figure 10 shows that the data are more scattered for this damage model compared with the model based on the plane of failure.

Another damage model is proposed that considers cyclic normal strain and cyclic shear strain rather than stress at the plane of failure. The damage model can be presented as,

$$D = \alpha(k\varepsilon_n + \gamma)_{mean} + \beta(k'\varepsilon_n + \gamma)_{amplitude} \qquad (23)$$

where α, β, k and k' are material constant and ε_n and γ are the strains at the critical plane of failure. The strains at the plane of failure were found with a similar transformation used for the stresses in equations (17) and (18). Fig. 11 shows that the above damage model predicts fatigue failure reasonably well for various loading conditions.

The multi-axial fatigue life data presented based on the above models show the data are somewhat scattered compared to the monotonic model. This could be related to the loading frequency and the visco-elastic properties of the composite material. It was observed during fatigue experiments that the temperature of the specimens substantially increased. A proper damage model should consider the effect of frequency and visco-elastic properties of composites. However, for low applied cyclic frequency, the effect of cyclic frequency could be ignored.

Fig. 5 Typical failure mode of the composite tube in a multiaxial fatigue experiment, showing failure is in the 45° plane to the specimen axis.

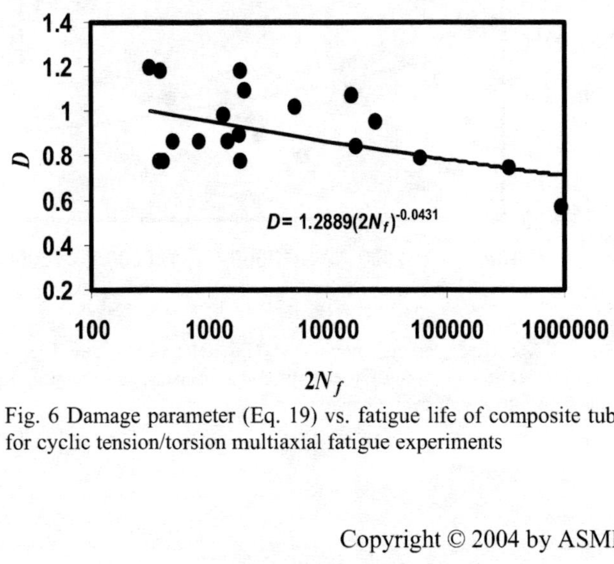

Fig. 6 Damage parameter (Eq. 19) vs. fatigue life of composite tube for cyclic tension/torsion multiaxial fatigue experiments

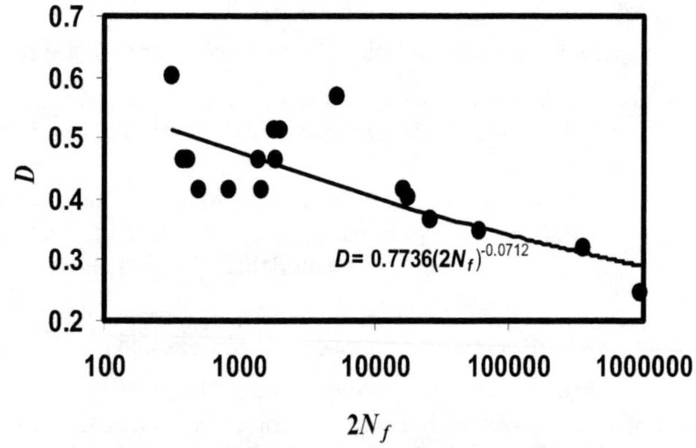

Fig. 7 Modified damage parameter (Eq. 20) vs. fatigue life of composite tube for cyclic tension/torsion multiaxial fatigue experiments, indicating a better correlation.

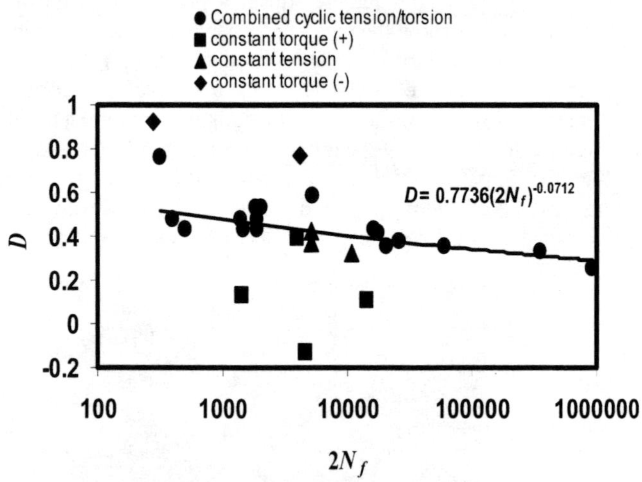

Fig. 8 Modified damage parameter (Eq. 20) vs. fatigue life of composite tube for various multiaxial fatigue experiments

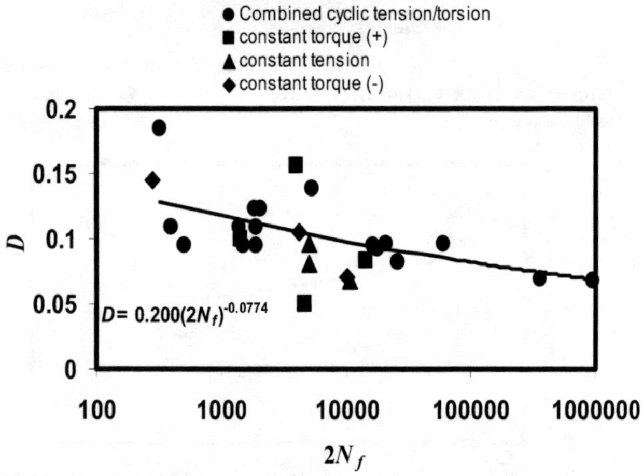

Fig. 9 Modified damage parameter (Eq. 21) vs. fatigue life for various multiaxial fatigue experiments, revised to account for constant torque and cyclic tension

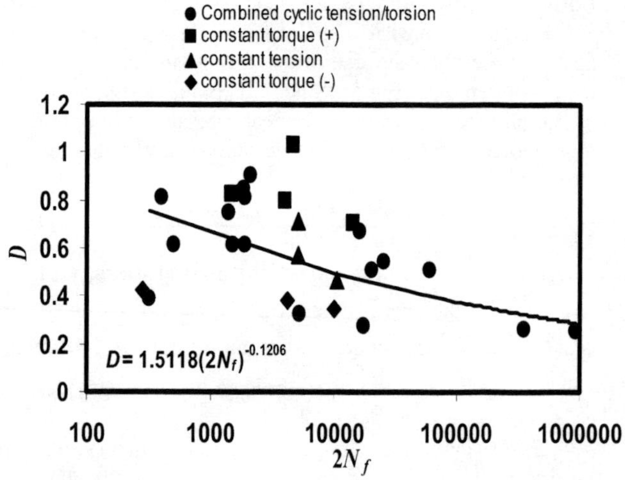

Fig. 10 Energy damage parameter (Eq. 22) vs. fatigue life for various multiaxial fatigue experiments

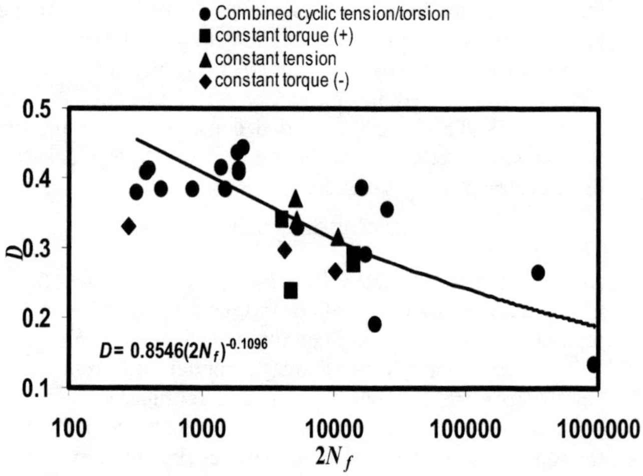

Fig. 11 Strain damage parameter (Eq. 23) vs. fatigue life for various multiaxial fatigue experiments

CONCLUSIONS

The failure observed during monotonic and fatigue failure experiments of E-glass/epoxy $[\pm 45^{\circ}]_4$ composite tubes under combined tension/torsion loading occurred in the outer laminate layer at 45° to the tube axis. A multiaxial monotonic failure criterion is proposed considering stresses at the plane of failure. The proposed failure criterion has an excellent correlation with the experimental data. Multiaxial fatigue failure models are proposed considering the mean and cyclic stresses at the plane of failure, the mean and cyclic strains at the plane of failure and the mean and cyclic energy. The experimental data has a good correlation with the proposed multi-axial fatigue failure models based on stress and strain at the plane of failure for various tension/torsion experiments. This is expected since there is a linear relationship between stress and strain seen during loading of the composite specimens. The proposed energy multi-axial fatigue failure model did not correlate as well to the experimental data as the other failure models. Future experiments should incorporate the temperature effect of loading frequency of the damage models. Other work is currently under investigation that utilizes the hysteresis curve of the multi-axial fatigue tests to predict the degradation of the strength of the composite.

REFERENCES

1. Found, M.S., 1985, "A Review of the Multiaxial Fatigue Testing of fiber Reinforced Plastics," Multiaxial Fatigue, ASTM STP 853, K.J. Miller and M.W. Brown, Eds., American Society of Testing Material, Philadelphia, pp. 318-395.
2. Radon, J.C. and Wachnicki, C.R., 1985, "Biaxial Fatigue of Glass Fiber Reinforced Polyester Resin," Multiaxial Fatigue, ASTM STP 853, K.J. Miller and M.W. Brown, Eds., American Society of Testing Material, Philadelphia, pp. 396-412.
3. Jones, D.L., Poulose, P.K. and Lebowitz, H., 1985, "Effect of Biaxial Loads on the Static and Fatigue Properties of Composite Materials," Multiaxial Fatigue, ASTM STP 853, K.J. Miller and M.W. Brown, Eds., American Society of Testing Material, Philadelphia, pp. 413-427.
4. Kaynak, C. and Mat, O., 2001, "Uniaxial Fatigue Behavior of Filament-Wound Glass/epoxy Composite Tubes," Composite Science and Technology, 61, pp. 1833-1840.
5. Krempl, E. and Niu, T., 1982, "Graphite/Epoxy $[\pm 45]_S$ Tubes. Their Static Axial and Shear Properties and Their Fatigue Behavior under Completely Reversed Load Controlled Loading," Journal of Composite Material, **16**, pp. 172-187.
6. Caprino, G., 2000, "Predicting Fatigue Life of Composite Laminates Subjected to Tension-Tension Fatigue," Journal of Composite Materials, **34**, pp. 1334-1355.
7. Lin, H., Nayeb-Hashemi, H., Pelloux, R.M.N. and Berg, C.A., 1992, "Cyclic Deformation and Anisotropic Constitutive Relations of Al-6061-T6 Under Biaxial Loading," Journal of Engineering Material and Technology, **114**, pp. 323-330.
8. Lin, H., Nayeb-Hashemi, H. and Pelloux, R.M.N., 1992, "Constitutive Relations and Fatigue Life Prediction for Anisotropic Al-6061-T6 Rods Under Biaxial Proportional Loadings," International Journal of Fatigue, pp. 249-259.
9. Lin, Hong, 1993, "Multiaxial Plasticity and Fatigue Life Prediction of Anisotropic Al-6061-T6," Ph.D. Thesis, Northeastern University, Boston, MA.
10. Socie, D.F., Waill, L.A. and Dittmer, D.F., 1985, "Biaxial Fatigue of Inconel 718 Including Mean Stress Effects," Multiaxial Fatigue, ASTM STP 853, K.J. Miller and M.W. Brown, Eds., American Society of Testing Material, Philadelphia, pp. 463-481.
11. Stephens, S.I., Fatemi, A., Stephens, R.R., and Fuchs H.O., 2001, *Metal Fatigue in Engineering, 2nd ed.*, John Wiley and Sons, Inc., New York, pp. 331-333,
12. Ellyin, F. and Golos, K., 1988, "Multiaxial Fatigue Damage Criterion," Journal of Engineering Material and Technology, **110**, pp. 63-68.
13. Ellyin. F, 1985, "Effect of Tensile-Mean-Strain on Plastic Strain Energy and Cyclic Response," Journal of Engineering Material and Technology, **107**, pp. 119-125.
14. Ellyin, F., Golos, K. and Xia, Z.,1991, "In-Phase and Out-of-Phase Multiaxial Fatigue," Journal of Engineering Materials and Technology, **113**, pp. 112-118.
15. Ellyin, F. and Xia, Z., 1991, "Non-proportional Multiaxial Cyclic Loading: Experiments and Constitute Modeling," Journal of applied Mechanics, **58**, pp. 317-325.
16. Ellyin F. and Kujawski, D., 1993, "Multiaxial Fatigue Criterion Including Mean-Stress Effect," Advances in Multiaxial Fatigue, ASTM STP 1191, D.L. McDowell and R. Ellis, Eds., American Society of Testing Materials, Philadelphia, pp. 55-66.
17. Leese, G.E. and Morrow, J., 1985 "Low Cycle Fatigue Properties of a 1045 Steel in Torsion," Multiaxial Fatigue, ASTM STP 853, K.J. Miller and M.W. Brown, Eds., American Society of Testing Material, Philadelphia, pp. 482-496.
18. Bannantine, J.A., Comer, J.J. and Handrock, J.L., 1990, *Fundamentals of Metal Fatigue Analysis*, Prentice Hall, Inc., Englewood Cliffs, N.J., pp. 69-70.
19. Nayeb-Hashemi, H., Cohen, M.D., Zotos, J. and Poormand, R., 1986, "Ultrasonic Characteristics of Graphite/Epoxy Composite Material Subject to Fatigue and Impacts," Journal of Nondestructive Evaluation, **5**, pp. 119-131.
20. Richardson, D., McLennan, M., Anderson, G., Macon, D. and Batista-Rodriguez, A., 2003, "Multiaxial Temperature- and Time-Dependent Failure Model," NASA Tech Briefs.

AUTHOR INDEX

PVP-Vol. 492
Recent Advances in Solids and Structures – 2004

Book Number: H01303